THE SOURCES OF INNOVATION

Eric von Hippel

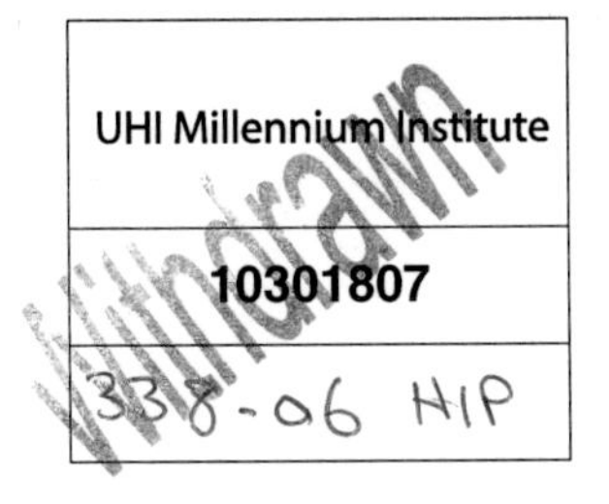

New York Oxford
Oxford University Press

Oxford University Press

Oxford New York
Athens Auckland Bangkok Bombay
Calcutta Cape Town Dar es Salaam Delhi
Florence Hong Kong Istanbul Karachi
Kuala Lumpur Madras Madrid Melbourne
Mexico City Nairobi Paris Singapore
Taipei Tokyo Toronto

and associated companies in
Berlin Ibadan

First published in 1988 by Eric von Hippel

First issued as an Oxford University Press paperback, 1995

Oxford is a registered trademark of Oxford University Press

Library of Congress Cataloging-in-Publication Data
Hippel, Eric von.
The sources of innovation.
Includes index.
1. Technological innovations—Economic aspects.
I. Title.
HC79.T4H56 1988 338′.06 86-28620
ISBN 0-19-504085-6
ISBN 0-19-509422-0 (Pbk.)

Figures 2.1 and 2.2 are reprinted from Eric von Hippel, "The Dominant Role of Users in the Scientific Instrument Innovation Process," *Research Policy,* July 1976, Vol. 5, No. 3.

9 8 7 6 5 4

Printed in the United States of America
on acid-free paper

Dedicated to:

ARvH

DvH

JRJ

ACKNOWLEDGEMENTS

The research reported on in this book spans a period of twelve years. In that time I have been helped by many colleagues, students, innovators, and research sponsors. I have striven to make the research and this book worthy of the generous help I have been given.

I would like to thank Thomas Allen, Anne Carter, Zvi Griliches, Ken-ichi Imai, Ralph Katz, Edwin Mansfield, Richard Nelson, Ikujiro Nonaka, Ariel Pakes, Richard Rosenbloom, and Roy Rothwell for giving me many valuable comments as the research proceeded.

I have been fortunate to have had a number of very talented visiting scholars and graduate students join me in research and discussion over the years. Especially prominent among these are: John Becker, Alan Berger, Julian Boyden, Alan Drane, Abbie Griffin, David Israel-Rosen, Andrew Juhasz, Toshihiro Kanai, Susumu Kurokawa, Walter Lehmann, Howard Levine, William Lionetta, Gordon Low, Richard Orr, Barbara Poggiali, Kiyonori Sakakibara, Stephen Schrader, Frank Spital, Heidi Sykes-Gomez, Pieter VanderWerf, and Walter Yorsz. All contributed greatly to the substance of the work and to the sheer fun of research.

Over the years my students and I have interviewed hundreds of people. Some were the developers of the important innovations we studied, while others had less direct knowledge. Many loaned us materials and all strove to help us to accurately understand their industries and their innovation-related experiences. Thanks to all.

The research I report on here would not have been possible without funding from the Division of Policy Research and Analysis of the National Science Foundation. Alden Bean, Miles Boylan, Andrew Pettifor, Rolf Piekharz, Eleanor Thomas, and anonymous peer reviewers supported my work over the years with a series of grants, despite budgets for extramural research that sometimes dropped perilously near zero.

Finally, I would like to thank Jessie Janjigian, who edited my manuscript and tried, with partial success, to teach me that respectable sentences can be less than a paragraph long, and that they need not include dashes—for emphasis.

Cambridge, Massachusetts EvH
October 1987

CONTENTS

THE SOURCES OF INNOVATION

1

Overview

It has long been assumed that product innovations are typically developed by product manufacturers. Because this assumption deals with the basic matter of who the innovator *is,* it has inevitably had a major impact on innovation-related research, on firms' management of research and development, and on government innovation policy . However, it now appears that this basic assumption is often wrong.

In this book I begin by presenting a series of studies showing that the sources of innovation vary greatly. In some fields, innovation users develop most innovations. In others, suppliers of innovation-related components and materials are the typical sources of innovation. In still other fields, conventional wisdom holds and product manufacturers are indeed the typical innovators. Next, I explore why this variation in the functional sources of innovation occurs and how it might be predicted. Finally, I propose and test some implications of replacing a manufacturer-as-innovator assumption with a view of the innovation process as predictably distributed across users, manufacturers, suppliers, and others.

The Functional Source of Innovation

Most of the studies in this book use a variable that I call the functional source of innovation. This involves categorizing firms and individuals in terms of the *functional* relationship through which they derive benefit from a given product, process, or service innovation. Do they benefit from using it? They are users. Do they benefit from manufacturing it? They are manufacturers. Do they benefit from supplying components or materials necessary to build or use the innovation? They are suppliers. Thus, airline firms are users of aircraft because the benefit they derive from existing types of aircraft—and the benefit they would expect to derive from innovative aircraft as well—are derived from use. In contrast, aircraft manufacturers benefit from selling aircraft, and

TABLE 1–1. Summary of Functional Source of Innovation Data

Innovation Type Sampled	*Innovation Developed by*					
	User	*Manufacturer*	*Supplier*	*Other*	*NA*[a] *(n)*	*Total (n)*
Scientific instruments	77%	23%	0%	0%	17	111
Semiconductor and printed circuit board process	67	21	0	12	6	49
Pultrusion process	90	10	0	0	0	10
Tractor shovel–related	6	94	0	0	0	16
Engineering plastics	10	90	0	0	0	5
Plastics additives	8	92	0	0	4	16
Industrial gas–using	42	17	33	8	0	12
Thermoplastics–using	43	14	36	7	0	14
Wire termination equipment	11	33	56	0	2	20

[a]NA = number of cases for which data item coded in this table is not available. (NA cases excluded from calculations of percentages in table.)

they would expect to benefit from an innovative airplane product by increasing their sales and/or profits.

Of course, the functional role of an individual or firm is not fixed; it depends instead on the particular innovation being examined. Boeing is a manufacturer of aircraft, but it is also a user of machine tools. If we were examining innovations in aircraft, we would consider Boeing to have the functional role of manufacturer in that context. But if we were considering innovations in metal-forming machinery, that same firm would be categorized as a user.

Many functional relationships can exist between innovator and innovation in addition to user, supplier, and manufacturer. For example, firms and individuals can benefit from innovations as innovation distributors, insurers, and so forth. As we will see later in this book, any functional class is a potential source of innovation under appropriate conditions.

Variations in the Source of Innovation

Novel ways of categorizing innovators are only interesting if they open the way to new insight. The first clue that the functional source of innovation is a potentially exciting way to categorize innovators comes with the discovery that the source of innovation differs very significantly between categories of innovation. Consider the several categories of innovation my students and I have studied in detail over the past several years (Table 1–1). In each study summarized in Table 1–1 the innovator is defined as the individual or firm that first develops an innovation to a useful state, as proven by documented, useful output.

Note the really striking variations in the functional source of innovation between the several innovation categories studied. Major product innovations in some fields, such as scientific instruments, are almost always developed by product users. In sharp contrast, product manufacturers are the developers of most of the important innovations in some other fields, and suppliers in still others (chapters 2 and 3).

An Economic Explanation

The observation that the functional source of innovation can vary is interesting in itself. But if we can understand the cause(s) of such variation, we may be able to predict and manage the innovation process much better.

There are many factors that influence the functional source of innovation. But we need not necessarily understand all of these in order to understand this variable usefully well and to predict the sources of innovation usefully often. As the reader will see, I propose that analysis of the temporary profits ("economic rents") expected by potential innovators can by itself allow us to predict the functional source of innovation usefully often (chapter 4).

This basic idea will certainly not be a surprise to economists. If it is to be useful in this context, however, certain preconditions must be met,* and expectations of innovation-related profits must differ significantly between firms holding different functional relationships to a given innovation opportunity. Since little is known about how firms formulate their expectations of profit from innovation, I have explored this matter in several detailed case investigations (chapter 5).

In all cases studied, it did appear that innovating firms could reasonably anticipate higher profits than noninnovating firms. The reasons for such differences varied from industry to industry. Interesting hints of general underlying principles did emerge, however, and sometimes these were related to the functional relationship between innovator and innovation. For example, users often had an advantage over other types of potential innovator with respect to protecting process equipment innovations from imitators. (Users often can profit from such an innovation while keeping it hidden behind their factory walls as a trade secret. This option is seldom available to manufacturers and others, who typically must reveal an innovation to potential adopters if they hope to profit from it.)

Understanding the Distributed Innovation Process: Know-how Trading Between Rivals

Once we understand in a general way why the functional sources of innovation may vary, we can graduate to exploring the phenomenon in greater

*I discuss and test conditions later; two, however, may convey the flavor. For an economic model of the functional source of innovation to predict accurately, it is necessary that potential innovators (1) not be able to shift functional roles easily and (2) benefit from their innovations by exploiting them themselves rather than by licensing them to others.

detail. Are there general strategies and rules that underlie how expectations of economic rents are formed and distributed across users, manufacturers, suppliers, and others? If so, we may gain a more general ability to predict how innovations will be distributed among these several functional categories of firm.

It is not clear a priori that useful rules for generating or predicting innovation strategies will exist: Such strategies are themselves a form of innovation, and one may not be able to describe the possibilities in terms of underlying components or rules. The only way to find out, I think, is by field investigation. I have undertaken one such investigation to date and have found an interesting phenomenon—informal know-how trading—that seems to me to have the characteristics of a generally applicable component for innovation strategies (chapter 6).

Informal know-how trading is essentially a pattern of informal cooperative R & D. It involves routine and informal trading of proprietary information between engineers working at different firms—sometimes direct rivals. (Know-how is the accumulated practical skill or expertise that allows one to do something smoothly and efficiently, in this instance the know-how of engineers who develop a firm's products and develop and operate its processes. Firms often consider a significant portion of such know-how proprietary and protect it as a trade secret.) Know-how trading exists in a number of industries my students and I have studied, and it seems to me to be an important phenomenon.

When I model informal know-how trading in terms of its effects on innovation-related profits, I find that one can predict when this behavior will or will not increase the expected profits of innovating firms. I propose that know-how trading between rivals is a general and significant mechanism that innovators can use to share (or avoid sharing) innovation-related costs and profits with rivals. As such, it is one of the tools we can develop and explore as we seek to understand the distributed innovation process.

Managing the Distributed Innovation Process: Predicting and Shifting the Sources of Innovation

Even though our understanding of the distributed innovation process is at an early stage, we should be able to get managerially useful results from it now. Indeed, it would be risky to not subject this work to the discipline of real-world experiment and learning by doing.

Since I have argued that variations in the sources of innovation are caused to a significant degree by variations in potential innovators' expectations of innovation-related profits, two managerially useful things should be possible. First, by understanding how expected innovation profits are distributed, we may be able to predict the likely source of innovation. Second, by changing the distribution of such profit expectations, we may be able to shift the likely source of innovation. If both of these fundamental things can be done, we would be well along the way to learning how to manage a distributed innovation process.

My colleague Glen Urban and I worked together to test the possibility of predicting the sources of a subset of user innovations: those having the potential to become commercially successful products in the general marketplace. (Not all user innovations have this characteristic. A user will innovate if it sees an in-house benefit from doing so and typically does not consider whether other users have similar needs. In contrast, a manufacturer typically requires that many users have similar needs if it is to succeed in the marketplace with a responsive product.)

The particular context of our test (chapter 8) was the rapidly evolving field of computer-aided-design equipment used to lay out printed circuit boards (PC-CAD).[1] Here we found that we could identify a subset of users that we termed lead users. We found innovation activity concentrated within this group as predicted: 87% of respondents in the lead user group built their own PC-CAD system versus only 1% of nonlead users. We also found that product concepts based on these lead user innovations were preferred by all users and therefore had commercial promise from the point of view of PC-CAD product manufacturers. This result suggests, by example, that prediction of sources of commercially promising innovation may be possible under practical, real-world conditions.

My colleague Stan Finkelstein and I tested the possibility of shifting the source of innovation in the field of automated clinical chemistry analyzer equipment (chapter 7).[2] Differences in clinical chemistry analyzer product designs were found that made some analyzer brands more expensive for innovating users to modify than others. If innovating users were seeking to maximize innovation benefit, we hypothesized that there should be more user innovation activity focused on the economical-to-modify analyzers—for experiments requiring equipment modification.

We tested this hypothesis in several ways and found it to be supported. We concluded that managers may sometimes be able to shift the sources of innovation affecting products of interest to them by manipulating variables under their control, such as product design.

Implications for Innovation Research

When a model fits reality well, data fall easily and naturally into the patterns predicted. I have been repeatedly struck by the clear, strong patterns that can be observed in the data that my students and I have collected on the functional sources of innovation. I hope that this aspect of the findings will not escape the eye of researchers potentially interested in exploring the functionally distributed innovation process.

Can we use the strong patterns identified in the functional sources of innovation to build a better understanding of the way innovation-related profits are captured? It seems to me to be important to do so: The nature and effectiveness of strategies for capturing innovation-related profits probably have as great an impact on economic behavior as do considerations of transaction costs or economies of scale, yet we know much less about them.

As an example of a strong pattern in the functional source of innovation data worth exploring, consider that my hypothesis simply states that innovations will be developed by those who expect a return they find attractive. But the data show that innovations of a specific type are typically developed by firms that expect the *most* attractive return. Can we build from this to show that those expecting the most attractive returns in an innovation category will invest more and eventually drive out all others over time? If so, we will greatly improve our ability to understand and predict the sources of innovation on the basis of innovation-related profits.

As an example of how a better understanding of real-world patterns in innovation-related profits may help us understand a range of economic issues, consider the matter of why firms specialize. Current explanations of this phenomenon focus on consideration of maximizing economy in production. But in the instance of process equipment, decisions by users to develop their own equipment do not appear to me to be motivated by such make-or-buy savings. Instead, innovating users appear to be motivated by considerations of increased profits they may obtain by having better equipment than that available to competitors. That is, they seem to be motivated primarily by considerations of innovation-related rents.

The research my students and I have carried out to date has primarily focused on product and process categories in which innovator firms have developed innovations on their own and have had only a single functional role with respect to those innovations. (An innovator was typically a user *or* a manufacturer, but not both.) The world clearly has more complex cases in it. In some fields of innovation, firms may customarily join with others to develop innovations cooperatively. In other fields firms may typically be vertically integrated or for some other reason contain multiple functional roles within the same organization. These more complex patterns should be studied. Possibly, but not certainly, we will be able to understand them in terms of the same principles found operating in simpler cases.

In the hope that some colleagues will find further research on the functional sources of innovation intriguing, I provide case materials (appendix). These may serve some as a convenient source of initial data.

Implications for Innovation Management

Innovation managers will, I think, find much of practical use in the research I explore in this book. The fact that the sources of innovation can differ has major consequences for innovation managers, both with respect to the organization of R & D and marketing and to management tools (chapter 9).

Firms organize and staff their innovation-related activities based on their assumptions regarding the sources of innovation. Currently, I find that most firms organize around the conventional assumption that new products are—or should be—developed by the firm that will manufacture them for commercial sale. This leads manufacturers to form R & D departments capable of fulfilling the entire job of new product development in-house and to organize market research departments designed to search for needs instead of innovations.

Indeed, if a manufacturer depends on in-house development of innovations for its new products, then such arrangements can serve well. But if users, suppliers, or others are the typical sources of innovation prototypes that a firm may wish to analyze and possibly develop, then these same arrangements can be dysfunctional. (For example, one cannot expect a firm's R & D group to be interested in user prototypes if its engineers have been trained and motivated to undertake the entire product development themselves.) Once the actual source of innovation is understood, the nature of needed modifications to firms' related organizational arrangements can be addressed.

New sources of innovation demand new management tools as well as new organization. Marketing research methods traditionally used to seek out and analyze user needs must be modified if they are to be effective for seeking out prototype products users may have developed. Similarly, tools for analyzing and possibly shifting the functional sources of innovation are not in firms' current management inventory and must be developed.

Early versions of needed tools will be found in this book. Obviously, much more work must be done. But I urge that innovative managers not wait for better tools and experiment now. Where patterns in the functional sources of innovation are strong, managers with a good understanding of their industries can get useful results by combining the basic concepts presented in this book with their own rich insights—and they should not be reluctant to do this.

Implications for Innovation Policy

Policymakers will find this research on the distributed innovation process interesting for many of the same reasons that managers will: Attempts to direct or enhance innovation must be based on an accurate understanding of the sources of innovation.

As was the case for innovation managers, government policymakers need new tools to measure and perhaps influence a functionally distributed innovation process, and these have not yet been developed. Pending the development of such tools, however, much can be done simply through an understanding that the innovation process can be a functionally distributed one.

As an illustration, consider the current concern of U.S. policymakers that the products of U.S. semiconductor process equipment firms are falling behind the leading edge. The conventional assessment of this problem is that these firms should somehow be strengthened and helped to innovate so that U.S. semiconductor equipment users (makers of semiconductors) will not also fall behind. But investigation shows (Table 1-1) that most process equipment innovations in this field are, in fact, developed by equipment *users*. Therefore, the causality is probably reversed: U.S. equipment builders are falling behind because the U.S. user community they deal with is falling behind. If this is so, the policy prescription should change: Perhaps U.S. equipment builders can best be helped by helping U.S. equipment users to innovate at the leading edge once more (chapter 9).

The elements in the example I have just described can clearly be seen as components in a distributed innovation process that interact in a systemlike manner. Eventually, I hope we will understand such systems well enough to have a ready taxonomy of moves, countermoves, and stable states. But even our present understanding of the functionally distributed innovation process can, in my view, help us to advance innovation research, practice, and policy-making.

Notes

1. Glen L. Urban and Eric von Hippel, "Lead User Analyses for the Development of New Industrial Products" (MIT Sloan School of Management Working Paper No. 1797–86) (Cambridge, Mass., June 1986), and *Management Science* (forthcoming).
2. Eric von Hippel and Stan N. Finkelstein, "Analysis of Innovation in Automated Clinical Chemistry Analyzers," *Science & Public Policy* 6, no. 1 (February 1979): 24–37.

2

Users as Innovators

In this chapter I begin by exploring who actually develops novel, commercially successful scientific instruments. Then I explore the actual sources of innovation in two major classes of process equipment used by the electronics industry. In both of these areas, I find that the innovators are most often users.

The discovery that users are innovators in at least some important categories of innovation propels us into the first major question I examine in this book: Who actually develops the vast array of new products, process equipment, and services introduced into the marketplace? The answer is clearly important: An accurate understanding of the source of innovation is fundamental to both innovation research and innovation management.

The Sources of Scientific Instrument Innovations

Scientific instruments are tools used by scientists and others to collect and analyze data. My study of scientific instrument innovations focuses on four important instrument types: the gas chromatograph, the nuclear magnetic resonance spectrometer, the ultraviolet spectrophotometer, and the transmission electron microscope. Each of these instrument types was, and is, very important to science.[1]*

*The gas chromatograph was a revolutionary improvement over previous wet chemistry methods used to identify chemical unknowns. Analyses that formerly took years to do or that could not be done at all prior to the innovation could now often be done in hours with gas chromatography. The nuclear magnetic resonance spectrometer (lately applied to medical research but initially used by chemists) opened an entirely new approach—the analysis of nuclear magnetic moments—to the determination of molecular structures. The ultraviolet spectrometer made analysis of materials by means of their ultraviolet spectra (a very useful research tool) easily achievable. The transmission electron microscope allowed researchers for the first time to create images of objects down to a resolution unit of approximately one angstrom (Å), far better than could be achieved by any optical microscope.

TABLE 2–1. Scientific Instrument Sample Composition

Instrument Type	*Improvements*			
	First-of-Type	*Major*	*Minor*	*Total*
Gas chromatograph	1	11	0	12
Nuclear magnetic resonance spectrometer	1	14	0	15
Ultraviolet absorption spectrophotometer	1	5	0	6
Transmission electron microscope	1	14	63	78
TOTAL	4	44	63	111

My innovation sample for each of the four instrument families included the initial, first-of-type device as it was first commercialized *and* the many commercially successful major and minor "improvement" innovations that enhanced the performance of that basic device over the succeeding 20 or more years.

The sample structure, shown in Table 2–1, might initially seem rather odd. Why focus on the innovations that improved just four types of scientific instrument? After all, many types of scientific instrument exist[2] and perhaps the generalizability of results might be better served by a random sampling from the whole field? Focusing on a few instrument types in depth, however, offers several advantages.

First, by examining successive innovations affecting a given instrument type, variables such as the nature of the market and industry structure, which might affect the sources of innovation we observe, can be controlled for. Second, a sample that follows the evolution of a few products over 20 or more years allows us a longitudinal view of the sources of innovation. Any major changes in the functional sources of innovation that may occur over time should be visible. Finally, an instrument type such as those examined here typically represents a product line from a manufacturing firm's viewpoint. Therefore, patterns of innovation that we observe in our samples are similar to those a manufacturer would have to face and deal with in the real world.

Methods

To guard against enthusiasm coloring my findings, I made my criterion for determining the source of an innovation objectively codable. I defined an *innovator* as the firm or individual that first developed a scientific instrument innovation to a state proved functionally useful, as indicated by the publication of data generated by it in a scientific journal.

My next task was to identify a sample of major and minor improvement innovations for each of the four instruments to be studied. This was done by, first, identifying users and manufacturer personnel expert in each instrument type.[3] Then, to identify major improvement innovations, each expert was asked to identify improvements developed after the basic innovation that provided a significant improvement in instrument performance relative to best preexisting practice.[4] The experts turned out to have quite uniform views. Either almost everyone contacted agreed that an innovation was of major functional utility—in which case it was included—or almost no one did, except the proposer—in which case it was rejected.

Minor improvement innovations were identified for the electron microscope only.[5] To generate a sample of these, the set of experts first listed all the innovations they could think of that had produced *any* improvement to any aspect of electron microscope performance and that had been commercialized. I then augmented this initial list by a scan of the catalogs of microscope manufacturers and microscope accessory and supply houses to identify any innovative features, accessories, specimen preparation equipment, and so on, that met the same criterion.

The samples of first-of-type and major improvement innovations that were identified by these procedures are listed in Table 2–2.

Samples in hand, I next faced a rather daunting data collection task. I wanted to understand the details of over 100 highly technical innovations and their histories. To accomplish the task I evolved a pattern that has served well during a number of studies. I set aside a summer and, with the aid of National Science Foundation (NSF) funding, recruited several excellent, technically trained MIT master's candidates to work with me. We all worked together in a large office, collecting data through telephone calls, library work, and field trips according to a standard data collection guide. Frequent comparing of notes and joint work (with breaks for noontime volleyball and chess games) kept our data to a high standard of reliability. (Additional discussion of data collection methods will be found in the appendix, along with detailed innovation case histories.)

The Sources of Innovation

As my students and I worked over the summer, we began to see that there was a clear answer to our question regarding the source of innovation in the field of scientific instruments. As can be seen in Table 2–3, it emerged that users were the developers of fully 77% of all the innovations we studied. And, as can be seen in Table 2–4, this pattern was uniformly present in all four instrument families studied.

Some sample members were not clearly independent: Several innovations were sometimes attributed to a single innovating user or manufacturer.[6] But, as is shown in Table 2–5, the finding of user innovation is not affected by this: A subsample that excludes all but the first case, chronologically, in which a particular user or firm plays a role shows the same pattern of innovation as the total sample. Employment of other decision rules in this test (e.g., the exclusion of all but the last case in which a given firm or user plays a role) produces the same outcome.

TABLE 2–2. Sample of Major Scientific Instrument Innovations

First-of-type: Gas chromatograph (GC)	
Major improvement innovations	
Temperature programming	Flame ionization detector
Capillary column	Mass spectrograph detector
Silanization of column support material	Gas sampling valve with loop
Thermal conductivity detector	Process control chromatography
Argon ionization detector	Preparative gas chromatography
Electron capture detector	
First-of-type: Nuclear magnetic resonance (NMR) spectrometer	
Major improvement innovations	
Spinning of NMR sample	Pulsed NMR spectrometer
Fourier transform/pulsed NMR	Heteronuclear spin decoupling
Homonuclear spin decoupling	Frequency synthesizer
Superconducting solenoids	Shim coils
Primas polecaps	T_1rho
Field frequency lock	Electronic integrator
Pulsed field gradient accessory	Proton-enhanced nuclear induction spectroscopy
Multinuclei probe	
First-of-type: Ultraviolet (UV) spectrophotometer	
Major improvement innovations	
Direct-coupled chart recorder	Automatic double beam
Automatic scanning	Double monochrometer
Reflection grating	
First-of-type: Transmission electron microscope (TEM)	
Major improvement innovations	
Pointed filaments	Three-stage magnification
Telefocus electron gun	Scaled-up objective pole piece
Double condenser lens	Goniometer specimen stage
Correction of astigmatism in objective lens	Cold-specimen stage
Well-regulated high-voltage power supplies	High-temperature specimen stage
Well-regulated lens power supply	Biased electron gun
Rubber gasket sealing of vacuum system	Out-of-gap objective lens

Recall that my measure of the source of innovation is based on who *first* developed a later-commercialized scientific instrument innovation. When users were found to be first, I termed them *the* innovators. But is it possible that in such cases manufacturers were also innovators, developing the same innovations independently? It seemed implausible, but I checked.

On the basis of two types of evidence, it appears that users who are first to innovate are indeed *the* innovators. First, most manufacturers who commercialize innovations initially developed by users *say* that their commercial product is based on the earlier, user-developed device. Second, as Table 2–6 shows, 78% of the instruments commercialized by scientific instrument manufacturers display the same underlying technical operating principles as their

TABLE 2–3. Source of Scientific Instrument Innovations by Innovation Significance

Innovation Significance	% User Developed	Innovation Developed by			
		User	Manufacturer	NA	Total
First-of-type	100%	4	0	0	4
Major improvement	82	36	8	0	44
Minor improvement	70	32	14	17	63
TOTAL	77	72	22	17	111

user prototype predecessors. This would be exceedingly unlikely to occur if users and manufacturers were engaged in parallel but independent research efforts.[7]

Three abbreviated case histories can convey a good feeling for the innovation patterns found in scientific instruments. The first is an example of a user-developed major improvement innovation; the second is an example of a manufacturer-developed major innovation; the third is an example of a minor improvement innovation developed by a scientific instrument user.

Case Outline 1. A user-developed major improvement innovation: spinning of a nuclear magnetic resonance sample.

Samples placed in a nuclear magnetic resonance spectrometer are subjected to a strong magnetic field. From a theoretical understanding of the nuclear magnetic resonance phenomenon, it was known by both nuclear magnetic resonance spectrometer users and personnel of the then-only manufacturer of nuclear magnetic resonance equipment (Varian Associates, Palo Alto, California) that increased homogeneity of that magnetic field would allow nuclear magnetic resonance equipment to produce more detailed spectra. Felix Bloch, Professor of Physics at Stanford University and the original discoverer of the nuclear magnetic resonance phenomenon, suggested that one could improve

TABLE 2–4. Source of Innovation by Type of Instrument

Major Improvement Innovations	% User Developed	Innovations Developed by			
		User	Manufacturer	NA	Total
Gas chromatograph	82%	9	2	0	11
Nuclear magnetic resonance spectrometer	79	11	3	0	14
Ultraviolet spectrophotometer	100	5	0	0	5
Transmission electron microscope	79	11	3	0	14
TOTAL	81	36	8	0	44

TABLE 2-5. A Subsample, Selected to Assure Independence, Shows Substantially the Same Pattern of User Innovation as Total Sample

Major Improvement Innovations	*% User Developed*	*Innovations Developed by*			
		User	*Manufacturer*	*NA*	*Total*
Gas chromatograph	86%	6	1	0	7
Nuclear magnetic resonance spectrometer	100	5	0	0	5
Ultraviolet spectrophotometer	100	2	0	0	2
Transmission electron microscope	83	5	1	0	6
TOTAL	90	18	2	0	20

the effective homogeneity of the field by rapidly spinning the sample in the field, thus averaging out some inhomogeneities. Two of Bloch's students, W. A. Anderson and J. T. Arnold, built a prototype spinner and experimentally demonstrated the predicted result. Both Bloch's suggestion and Anderson and Arnold's verification were published in the same issue of *Physical Review*.[8]

Varian engineers went to Bloch's laboratory, examined his prototype sample spinner, developed a commercial model, and introduced it into the market by December 1954. The connection between Bloch and Varian was so good and Varian's commercialization of the improvement so rapid that there was little time for other users to construct homebuilt spinners prior to that commercialization.

Case Outline 2. A manufacturer-developed major improvement innovation: a well-regulated, high-voltage power supply for transmission electron microscopes.

The first electron microscope and the first few precommercial replications used batteries connected in series to supply the high voltages they required. The major inconvenience associated with this solution can be readily imagined: voltages on the order of 80,000 v were required, and nearly 40,000

TABLE 2–6. Were the Operating Principles of the User's Design Replicated in the First Commercial Device?

Major Improvement Innovations	*%Yes*	*Yes*	*No*	*NA*	*Total*
Gas chromatograph	78%	7	2	0	9
Nuclear magnetic resonance spectrometer	82	9	2	0	11
Ultraviolet spectrophotometer	100	5	0	0	5
Transmission electron microscope	64	7	4	0	11
TOTAL	78	28	8	0	36

single wet-cell batteries had to be connected in series to provide this. A visitor to the laboratory of L. Marton, an early and outstanding experimenter in electron microscopy, recalls an entire room filled with batteries on floor-to-ceiling racks with a full-time technician employed to maintain them. An elaborate safety interlock system was in operation to insure that no one would walk in, touch something electrically live, and depart this mortal sphere. Floating over all was the strong stench of the sulfuric acid contents of the batteries. Clearly, not a happy solution to the high-voltage problem.

The first commercial electron microscope, built by Siemens of Germany in 1939, substituted a power supply for the batteries but could not make its output voltage as constant as could be done with batteries. This was a major problem because high stability in the high-voltage supply was a well-known prerequisite for achieving high resolution with an electron microscope.

When RCA decided to build an electron microscope, an RCA electrical engineer, Jack Vance, undertook to build a highly stable power supply and by several inventive means achieved a stability almost good enough to eliminate voltage stability as a constraint on the performance of a high-resolution microscope. This innovative power supply was commercialized in 1941 in RCA's first production microscope.

Case Outline 3. A user-developed minor improvement innovation: the self-cleaning electron beam aperture for electron microscopes.

Part of the electron optics system of an electron microscope is a pinhole-sized aperture through which the electron beam passes. After a period of microscope operation, this aperture tends to get contaminated with carbon. The carbon becomes electrically charged by the electron beam impinging on it; the charge in turn distorts the beam and degrades the microscope's optical performance. It was known that by heating the aperture one could boil off carbon deposits as rapidly as they formed and thus keep the aperture dynamically clean. Some microscope manufacturers had installed electrically heated apertures to perform this job, but these devices could not easily be retrofitted to existing microscopes.

In 1964 a microscope user at Harvard University gave a paper at the EMSA (Electron Microscope Society of America) in which he described his inventive solution to the problem. He simply replaced the conventional aperture with one made of gold foil. The gold foil was so thin that the impinging electron beam made it hot enough to induce dynamic cleaning. Since no external power sources were involved, this design could be easily retrofitted by microscope users.

C. W. French, owner of a business that specializes in selling ancillary equipment and supplies to electron microscopists, read the paper, talked to the author/inventor, and learned how to build the gold foil apertures. French first offered them for sale in 1964.

The User's Role in Innovation Diffusion

The innovating users in the case histories presented were researchers employed by universities. And, as we see in Table 2–7, this was generally true for my sample of user-developed innovations.

TABLE 2–7. Institutions Employing Innovative Users

Major Improvement Innovations	*University/ Institute*	*Private Manufacturing Firm*	*Self-employed*	*NA*	*Total*
Gas chromatograph	3	3	1	2	9
Nuclear magnetic resonance spectrometer	9	0	0	2	11
Ultraviolet spectrophotometer	4	1	0	0	5
Transmission electron microscope	10	0	0	1	11

Given that the innovating scientific instrument users were university scientists, we might expect them to be very active in speeding the diffusion of their innovations—and they were. First (as required by the mores of science), innovating users (researchers) published their research results *and* the details of any homebuilt apparatus used to attain them. Second, they typically also informed others of their innovations by presentations at conferences and visits to the laboratories of other scientists.

Information diffused by innovators regarding major innovations was rapidly picked up by other scientists or by commercializing firms. In the instance of major improvements to GC or NMR (the two areas where I looked into the matter) one of two types of diffusion occurred within a year after the initial publication by the original innovating user: Either (1) other scientists replicated the homebuilt device and also published papers involving its use (frequently the case) or (2) a commercial version was on the market (seldom the case). Both patterns are shown in Table 2–8.

In sum, we see that the role of the user—depicted schematically in Figure 2–1—was both very rich and central to the scientific instrument innovation process.

TABLE 2–8. When Instrument Manufacturers Did Not Commercialize User Innovations Quickly, Other Users Made Homebuilt Copies

	User time lag > 1 year				*Homebuilts present, time lag 1 year or < 1 year*			
Innovation	*%Yes*	*Yes*	*No*	*NA*	*%Yes*	*Yes*	*No*	*NA*
Gas chromatograph	100%	5	0	0	0%	0	3	1
Nuclear magnetic resonance spectrometer	100	8	0	1	0	0	1	1
TOTAL	100	13	0	1	0	0	4	2

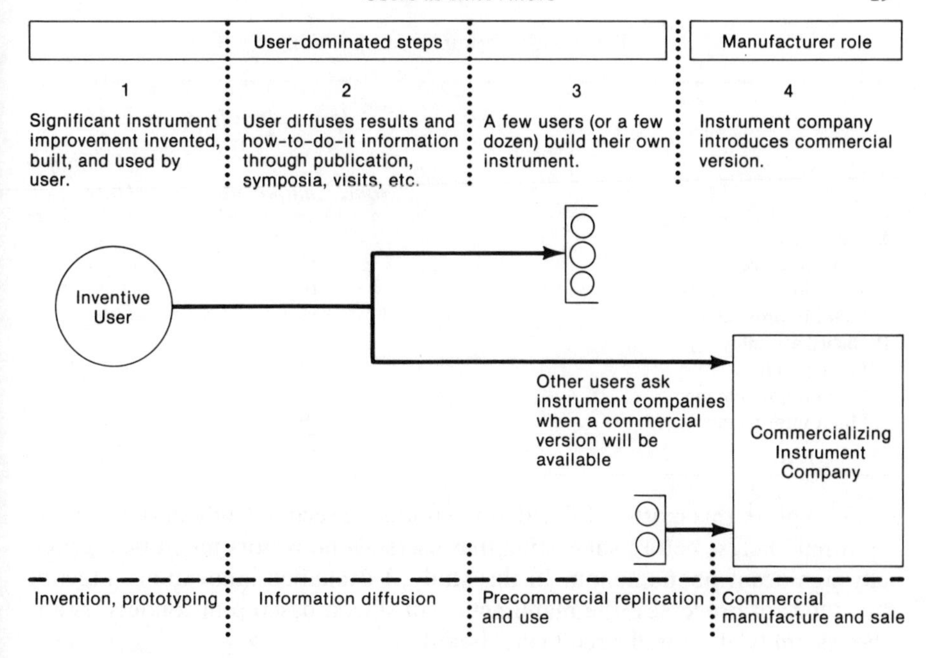

FIGURE 2–1. Typical Steps in the Development and Diffusion of a Scientific Instrument Innovation

Typically the innovative user:

- Perceived that an advance in instrumentation was required.
- Invented the instrument.
- Built a prototype.
- Proved the prototype's value by applying it.
- Diffused detailed information on both the value of the invention and on how the prototype device could be replicated.

In instances coded as user innovation, an instrument manufacturer entered the process only after all of the above events had transpired. Typically, the manufacturer then:

- Performed product engineering work on the user's device to improve its reliability and convenience of operation.
- manufactured, marketed, and sold the innovative product.

The Sources of Semiconductor and Printed Circuit Board Assembly Process Innovations

The study of scientific instruments I have just reviewed showed user innovation as typical in that field. But is this pattern unique to scientific instruments? After all, university scientists, the typical innovators in that field, are clearly not typical of the users of most products, processes, or services.

TABLE 2–9. Sample Composition

	Innovations Implemented by	
	Novel Equipment	*Novel Technique Only*
Semiconductor		
Initial practice	5	6
Major improvements	16	3
Minor improvements	11	0
PC board assembly		
Initial practice	2	2
Major improvements	6	0
Minor improvements	9	0
TOTAL	49	11

To explore this matter, I decided to conduct a second study in other, more "normal" fields, before suggesting that users-as-innovators might be a generally significant phenomenon. In this study, I examined innovations affecting two types of processes: the manufacture of silicon-based semiconductors and the assembly of printed circuit (PC) boards.*

Methods

Semiconductors and PC boards are, in common with most products, manufactured by means of a series of process steps. Thus, the process of manufacturing silicon-based semiconductors may start with a crystal-growing process step, followed by a step in which the crystal is sliced into thin circular wafers, and so forth.

My sample in this study consisted of the successive innovations that first established and then improved several such manufacturing process steps (see Table 2–9). Since the machinery used for a manufacturing process step often represents a product line for an equipment manufacturer, the resulting sample structure is similar to that used in the study of scientific instruments, and it shares its advantages.

The 60 innovations included in this study (listed in Table 2–10) were identified by means of a process involving several steps. I began by studying process flow sheets to identify the major process steps used to manufacture semiconductors and assemble printed circuit boards. Next, I selected some of these process steps[9] and identified the method used in the initial commercial practice of each (i.e., the first method used by any firm to manufacture products for sale rather

*Most electronic products today use printed circuit boards to link the electronic components they contain (integrated circuits, resistors, capacitors, etc.) into functioning circuits. The PC board itself resembles a plastic board or card. It is typically rectangular, it is less than 1/16-in. thick, and it measures a few inches on each side. Electronic components are mounted on one or both board surfaces, and thin metal paths that run on the surface of the board and/or within it interconnect the components into the desired electronic circuitry. Board manufacture here includes the manufacture of the basic board, component insertion, interconnection, and testing.

TABLE 2–10. Innovations Identified for Silicon Semiconductor and for Printed Circuit Board Subassembly Processing[a]

Major Process Step	*Initial Commercial Practice*	*Major Improvement*
Silicon semiconductor products		
1. Growth of single-silicon crystal[b]	Crystal puller	Resistance-heated crystal puller Dislocation-free crystal puller[c] Automatic diameter control
2. Wafer slicing	High-precision saws[d]	ID saw
3. Wafer polishing	Optical polishing equipment and technique[d]	Chemical/mechanical polishing (SiO_2)[d] Chemical/mechanical polishing (Cupric salts)[d]
4. Epitaxial processing (optional process step)	Pancake reactor	Horizontal reactor Barrel reactor
5. Oxidation	Not examined	
6. Resist coating	Wafer spinner	High acceleration wafer spinner 11 minor improvement innovations
7. Mask alignment and wafer exposure	Mask aligner	Split field optics aligner Automated mask aligner
8. Oxide etching	Not examined	
9. Silicon junction fabrication	Grown junction[c]	Diffused junction furnace Ion implantation accelerator
10. Metalization	Not examined	
11. Scribing and dicing	Jig and fixture[c]	Mechanical scriber and dicer Laser scriber and dicer
12. Mounting	Not examined	
13. Wire bonding	Solder bonding	Thermocompression bonding Ultrasonic bonding
14. Encapsulation	Not examined	
15. Mask graphics	Handcut rubylith patterns[c]	Optical-pattern generator Electron beam pattern generator
16. Mask reduction	Two-stage step and repeat reduction process	Not examined
Electronic subassembly manufacture		
1. Circuit fabrication	PC board[d] Wire wrapping (optional)	Not examined[d] Automated wire wrapping (optional)
2. Component insertion	Hand component insertion[c]	Single-component-per-station component insertion *X–y* table component insertion Numerically controlled-driven *x–y*-table component insertion Sequenced component insertion
3. Mass soldering	Dip solder[c]	Wave solder 9 minor improvement innovations
4. Assembly	Not examined	

[a]*Source*: Eric von Hippel, "The Dominant Role of the User in Semiconductor and Electronic Subassembly Process Innovation," *IEEE Transactions on Engineering Management* EM-24, no. 2 (May 1977), 64–65 © 1977 IEEE.

[b]Float zone refining and dislocation-free float zone refining offer an alternate silicon single crystal growing technology.

[c]This process innovation was embodied primarily in operator technique rather than in novel process equipment.

[d]The process machinery used in the initial commercial practice of this process step was commercially available and being used in other industries. Innovation work needed in these instances consisted simply of identifying the equipment as appropriate for the process step contemplated and/or redefining the process step specifications until they fitted the capabilities of that equipment.

TABLE 2–11. Sources of Process Machinery Innovations

	Innovation Developed by					
	% User	*User*	*Manu-facturer*	*Joint User/ Manu-facturer*	*NA*	*Total*
Semiconductor process						
Initial practice	100%	5	0	0	0	5
Major improvements	71	10	2	2	2	16
Minor improvements	56	5	3	1	2	11
PC board assembly						
Initial practice	100	2	0	0	0	2
Major improvements	40	2	2	1	1	6
Minor improvements	63	5	2	1	1	9
TOTAL	67	29	9	5	6	49

than for laboratory purposes). Then, using the same process of polling experts described earlier in the context of the scientific instrument study, I identified the major improvements that had been made to each process step over the following years. (A *major improvement* was defined as a change in equipment or technique that provided a significant improvement in process step performance relative to best preinnovation practice.) Finally (again following methods described earlier), I identified an exhaustive sample of minor process step improvements affecting one semiconductor and one PC board assembly process step. (*Minor improvement* innovations were defined as those that gave the user any improvement in any dimension important in processing such as cost reduction, increased speed, quality, consistency, and so on.[10])

As in the scientific instrument study, I defined an innovator as the firm or individual that first developed a sampled innovation to a state *proved* functionally useful. Here, proof of functional usefulness was documented use of the innovation in commercial production. All of the innovations selected for study were commercially successful, with commercial success being defined as near-universal adoption by process users in the few years following the innovation's debut. (Today, of course, many of the innovations have been supplanted by later improvements.)

Data collection methods used in this study are precisely the same as those used in the study of scientific instruments that were described earlier in this chapter.

The Sources of Innovation

As Table 2–11 shows, users developed all of the process machinery innovations involved in the initial commercial practice of a process step and more than 60% of the major and minor improvements to that machinery. (Conventional wisdom suggests that user-developed innovations are rare. But even if

TABLE 2–12. Process Innovations That Do Not Require Novel Equipment

	Innovations Implemented Through		
	Commercial Equipment[a]	*Technique Only*[b]	*Developed by*
Semiconductor process			
Initial practice	2	3	100% User
Major improvements	0	3	100% User
PC board assembly			
Initial practice	1	1	100% User
Major improvements	0	0	—
TOTAL	3	7	

[a]Identified in Table 2–10 by the superscript "d."
[b]Identified in Table 2–10 by the superscript "c."

we allow H_0 to be that users will develop 50% of the sampled innovations, $p <$ .02, our sample would yield the 67% user-developed innovations reported in Table 2–11.) Clearly, user innovation is not a phenomenon restricted to scientific instruments only.

In this second study we see a modest amount of joint user/manufacturer innovation activity (coded as user/manufacturer in Table 2-11). Also, we see users active in two types of innovations that I have not discussed before.

First, from Table 2-12 note that users developed all of the technique-only process innovations in the sample. (Such an innovation does not require any novel equipment for implementation. Rather, it involves modifying the way in which existing equipment is operated in order to make an improvement.)

Second, users were found to be the developers of all (three only) multistep process concepts I examined in this study. These are the important process concepts that underlie single process steps and give them meaning. For example, in the semiconductor industry the process steps listed as 5–10 in Table 2–10 are all steps in a photolithographic process that are intended to implement a larger process concept known as the planar process of semiconductor manufacture.

I did not explicitly collect a sample of this important type of innovation, but I did note three in the course of collecting data on the single-step innovations I was studying. The first of these was the planar process for manufacturing semiconductors just mentioned. It was developed by Fairchild Semiconductor, a process user.The product/process concept of building semiconductors on a silicon substrate rather than on germanium also affected many process steps, and it was developed by Bell Laboratories and Texas Instruments, both users of the process. Finally, the basic product/process concept of mounting electronic components on a plastic board that had electrical circuits printed on it (the basic concept of the PC board) was developed by the U.S. Signal Corps, a user, in 1948 as part of an effort to miniaturize military electronics.

TABLE 2–13. Patterns in Transfer of User Innovations to First Commercializing Equipment Manufacturers[a]

	How Frequently Observed		Lag Between User Innovation and First Commercial Equipment Sale	
Pattern Observed	*%*	*(n)*	*Mean Years*	*SD*
Multiple user/ manufacturer interactions	46%	11	3.7	1.3
No transfer found	25	6	1.8	0.4
User equipment order	21	5	1.0	1.3
User becomes manufacturer	8	2	4.0	0.0

[a]Total user-developed process machine innovations = 29; transfer pattern data NA = 5.

Diffusion of Innovations

Unlike the situation in scientific instrument innovations, users of process equipment innovations do not necessarily have an incentive to transfer what they know to an equipment manufacturer. In fact they might have an incentive to hide what they know to achieve a competitive advantage. (I will consider this matter in depth in chapter 6.) Therefore, it would be interesting to know how equipment manufacturers learned of the user process innovations I studied, and I looked into the matter.[11]

Details of the transfer process were typically not well documented or recalled by interviewees. However, I determined that the transfer of user-developed process equipment innovations to the first equipment manufacturing firm to produce them as a commercial product fell into one of four general patterns (see Table 2–13).

In order of the frequency with which these were observed:

1. Multiple interactions between the staffs of user firms and manufacturer firms made it impossible to isolate *the* events surrounding transfer. In these instances, several user firms had homemade versions of the innovation in-house at the time of transfer, and it was clear that a great deal of information was being passed around. A typical interviewee comment: "Everyone was talking about *x* user design at the time."
2. No transfer identified. Although a user was first to develop the equipment used commercially (and was coded as the innovator on this basis), no transfer process was identifiable retrospectively.
3. A user (not necessarily the initial innovating user) transferred the design of the innovation along with a purchase order for units produced to that design. The user's intent in these instances was to obtain an outside source of supply for the novel equipment.
4. An equipment user (not necessarily the innovating user) also adopted the role of equipment manufacturer and began to produce the innovation for sale to other user firms.

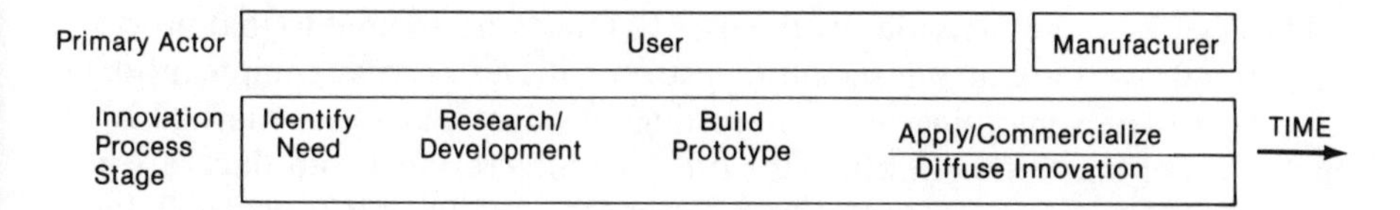

FIGURE 2–2. Steps Observed in the User Development of an Innovation

The User-Dominated Innovation Process

We have now found three innovation categories in which it is typically the product user, not the product manufacturer, who recognizes the need, solves the problem through an invention, builds a prototype, and proves the prototype's value in use. If we apply this finding to "stages" of the technical innovation process, we find—somewhat counterintuitively—that the locus of almost the entire innovation process is centered on the user. As is shown schematically in Figure 2–2, only commercial diffusion is carried out by the manufacturer.

This finding is at odds with conventional wisdom and with most of the prescriptive literature in the new product development process directed to manufacturers. That literature characteristically assumes that the *manufacturer* must find a need and fill it by executing the new product development stages in Figure 2–2.

It is perhaps natural to assume that most or all of the innovation process culminating in a new industrial good occurs within the commercializing firm. First, as we will see in chapter 3, the manufacturer *is* the usual innovator in many product and process categories. Second, a manufacturer's association with an innovation is usually much more public than that of users and others, and this can inadvertently reinforce the presumption (sometimes false) of manufacturer-as-innovator. Thus, very naturally, in the course of marketing an innovation, manufacturing firms may advertise "their" innovative device. These firms do not mean to imply that *they* invented, prototyped, and field tested the advertised innovation. But, in the absence of countervailing advertising by innovating users or other contributors to the innovative process (advertising they generally have no reason to engage in), it is easy to make the assumption.

Of course, some might feel that the data presented in this chapter are *not* evidence of user innovation and that the within-manufacturer "norm" applies. One might decide, for example, that the user-built prototype of an innovative instrument available to an instrument firm simply serves as a new product "need" that the firm (in the terminology of Figure 2–2) "identifies." It would then follow that the succeeding stages in Figure 2–2 also occur within the manufacturing firm. The "research and development" stage, for example, might consist of the engineering work manufacturer personnel devote to converting the user prototype into a commercial product.

Although one might make the argument outlined above, I myself find it rather thin and unproductive to do so. Essentially, the argument enshrines relatively minor activities within the manufacturer as the innovation process and relegates major activities by the user to the status of input to that process. If, instead, we look at the scientific instrument and process equipment data afresh, we see something very interesting: Product categories marked by a great deal of innovation in which the firms manufacturing the products are not necessarily innovative in and of themselves. Indeed, we might plausibly look at the manufacturers of these products as typically only providing the manufacturing function for an innovative set of user/customers.

This finding that nonmanufacturers may be the innovators in some industries certainly opens the way to an interesting new view of the innovation process. After all, accurate knowledge of who the innovator *is* is essential to much innovation research and practice.

Notes

1. National Research Council of the National Academy of Sciences, *Chemistry: Opportunities and Needs* (Washington, D.C.: National Academy of Sciences, 1965), 88.

2. U.S. Department of Commerce, Bureau of the Census, *Current Industrial Reports: Selected Instruments and Related Products* (MA38–B (80)–1 January 1982 SIC Code 38112) (Washington, D.C.: U.S. Government Printing Office, 1982).

3. The experts consulted were, on the manufacturer side, senior scientists and/or R & D managers who had a long-time (approximately 20 years) specialization in the instrument family at issue and whose companies have (or, in the case of electron microscopy, once had) a share of the market for that instrument family. The users consulted were interested in instrumentation and/or had made major contributions to it (as evidenced in scientific review articles of each field).

4. This decision rule excluded "me-too" innovations from the sample, including those that duplicated the successful performance increase of a previous innovation but by different technical means.

5. Much of the improvement in performance of a product or process can be the cumulative result of many minor, incremental innovations (see Samuel Hollander, *The Sources of Increased Efficiency: A Study of Du Pont Rayon Plants* [Cambridge, Mass.: MIT Press, 1965], 196; Kenneth E. Knight, "A Study of Technological Innovation: The Evolution of Digital Computers" [PhD diss., Carnegie Institute of Technology, Pittsburgh, Penn., 1963]). Therefore, I had wanted to identify samples of minor improvement innovations for all four types of scientific instruments being studied. As work proceeded, however, I only carried out this plan in the instance of the transmission electron microscope. Unfortunately, experience showed that participants could not recall events surrounding minor innovations very well or very reliably. The events had not seemed very significant at the time—indeed, they were not, they were minor—and the details had faded with time.

6. The community of users and manufacturers associated with each of the four instrument types I studied was small in the early days of each type. Therefore, as is reasonable, my data contain several instances in which more than one major innovation was invented by the same user or first commercialized by the same instrument firm. With respect to a single user developing more than one innovation: 2 GC innovations were developed by a single user, as were 3 NMR innovations, 2 UV innovations,

and 4 TEM innovations. With respect to a single manufacturing firm being first to commercialize more than one sample innovation, the 111 innovations in my sample were first commercialized by only 26 companies: 12 GC innovations, first commercialized by 8 companies; 15 NMR innovations, first commercialized by 3 companies; 6 UV innovations, first commercialized by 2 companies; 15 TEM basic and major improvement innovations, first commercialized by 6 companies; and 63 TEM minor innovations, first commercialized by a total of 7 companies.

7. The coding of this question involves some existence of technical judgment by the coder as no clear definitional boundary exists between the operating principles of an invention and its engineering embodiment. Perhaps I can best convey a feeling for the two categories by an illustration using Felix Bloch's sample spinning innovation described later in this chapter. The *concept* of achieving an effective increase in magnetic field homogeneity through the operating principle of microscopically spinning the sample can have many engineering embodiments by which one achieves the desired spin. Thus one company's embodiment may use an electric motor to spin a sample holder mounted on ball bearings; another might, in effect, make the sample holder into the rotor of a miniature air turbine, achieving both support and spin by means of a carefully designed flow of air around the holder.

8. F. Bloch, "Line-Narrowing by Macroscopic Motion," *Physical Review* 94, no. 2 (15 April 1954): 496–97; W. A. Anderson and J. T. Arnold, "A Line-Narrowing Experiment," *Physical Review* 94, no. 2 (15 April 1954): 497–98.

9. I originally planned to study all 21 major process steps identified in Table 2–10. Because of time limitations, however, only 14 were completed. These were not chosen for study randomly, but were chosen by no conscious system.

10. Innovations that offered major or minor increments in functional utility to users relative to previous best practice were identified independently for each process step studied (i.e., major improvements in component insertion equipment were identified by comparison with other component insertion equipment innovations only). This was done because improvements in the different types of equipment typically had an impact on various dimensions (precision, speed, reliability, and so on) not easily made commensurable.

11. Eric von Hippel, "Transferring Process Equipment Innovations from User-Innovators to Equipment Manufacturing Firms," *R&D Management* 8, no. 1 (October 1977): 13–22.

3

Variations in the Functional Source of Innovation

We have seen that users sometimes innovate. But do they always? Or does the functional source of innovation vary in some manner between users, manufacturers, suppliers, and others? To be able to answer these questions, we must have data on the sources of innovation characteristic of at least a few more types of innovation. Therefore, my students and I undertook the six brief studies I will describe in this chapter.

Each of the six studies examines a different type of innovation. The first four I will review consider innovation categories chosen to match my own areas of technical knowledge and those of the graduate students participating in our project. The choice of topics for the last two studies I will review was made on a different basis, which I will spell out when discussing them.

Each study uses identical methods, so that their results are commensurable, and each is tightly focused on a single issue: What are the sources of innovation? These studies find that the functional source of innovation differs strikingly across the several types of product and process innovation I have explored.

Users as Innovators: Pultrusion

Pultrusion is a valuable process for manufacturing fiber-reinforced plastic products of constant cross-section. It is well suited to the production of high-strength composites with reinforcement material that is aligned in known directions—just what is needed for demanding structural applications such as those in aerospace vehicles and sports equipment. Although pultrusion sales at the time of the study were relatively small, they had been growing rapidly and were expected to continue to do so.[1]

The pultrusion process is performed from start to finish on a single ma-

TABLE 3–1. Pultrusion Process Machine Innovations

Basic Innovation: Original batch pultrusion process	
	Major process machinery improvements
Intermittent pultrusion process	Tractor pullers
Tunnel oven cure	Cut-off saw
Continuous pultrusion process	Radio frequency augmented cure
	Tooling innovations
Preforming tooling	Improved dies
Hollow product tooling	

chine. The process starts when reinforcing material such as fiberglass is pulled simultaneously from a number of supply rolls and into a tank containing a liquid thermoset resin such as polyester. Strands of reinforcement material emerge from the tank thoroughly wetted with resin and then pass through preforming tooling that aligns and compacts the strands into the desired cross-section. The compacted bundle of glass and liquid resin is then pulled through a heated die where the resin is cured. Next the cured product moves through pullers, which are the source of the considerable mechanical force needed to draw reinforcing material and resin through the steps just described. Finally a saw cuts the continuously formed product into sections of the desired length.

The Sample

The basic pultrusion process was developed in the late 1940s. Since that time, there have been major improvements to pultrusion process machinery and to the resins and reinforcement materials used in the pultrusion process as well. The sample (Table 3–1) focuses on pultrusion process machinery only. It contains all machinery innovations that resulted in major improvements to the pultrusion process when judged relative to the best practices obtaining at the time each innovation was first commercialized.[2]

In addition, the sample contains three tooling innovations. These are not strictly part of a general-purpose pultrusion machine. Rather, they are accessories to the machine (sometimes also called jigs or fixtures) that are designed especially to aid in the manufacture of a particular product. (The ones in the sample are each useful for a large category of pultruded product. Each is described in detail in the appendix.)

All innovations studied were very successful and spread through much of the user community in the form of user-made and/or commercially produced equipment.

Findings: The Sources of Pultrusion Equipment Innovation

As can be seen in Table 3–2, almost all significant pultrusion process machinery innovations were developed by machine users (producers of pultruded products).

TABLE 3–2. Sources of Pultrusion Process Machinery Innovations

		Innovation Developed by			
Innovation Type	*% User*	*User*	*Manufacturer*	*NA*	*Total*
Original process	100%	1	0	0	1
Major improvements	86	5	1	0	6
Tooling innovations	100	3	0	0	3
TOTAL	90	9	1	0	10

This finding follows the pattern of earlier studies, but in this case the user-innovators appear much less technically sophisticated than those we encountered earlier. Innovating users of pultrusion process equipment were emphatically not high-tech firms. They were essentially job shops making fiberglass products by pultrusion and, often, by hand-layup methods as well. (Readers who remember repairing the bodywork on their cars as teenagers will have had firsthand experience with the hand-layup process. It simply involves wetting fiberglass fabric or roving with a liquid plastic, shaping the wet material as desired, then curing the plastic.)

These user firms had no formal R & D groups and, typically, no one with formal technical training in plastics or plastics fabrication. If an order came in for a part of novel shape, the foreman on the factory floor or one of the workers would make up a mold to aid in shaping it. If the first design worked, production would commence. If not, the mold and patterns of fiberglass layup would be tinkered with until they did work acceptably. The more parts needed of a given shape, the greater the effort put into making molds and other aids to speed the layup process.

Typically, important pultrusion innovations were triggered when a user firm received a large order for a part of uniform cross-section such as hundreds of feet of one structural shape or thousands of feet of rod to be used for fiberglass fishing poles. Faced with a massive task of this sort, a creative person on the factory floor was sometimes inspired to innovate, using an innate sense of engineering design and machine parts lying around the factory. In a few instances, these efforts resulted in process equipment innovations of general value.

Manufacturers as Innovators: The Tractor Shovel

The tractor shovel is a very useful machine often used in the construction industry. Initial conversations with experts in construction led us to suspect that users would in fact be innovators in tractor shovels. Everyone had a story to tell about a construction firm that, facing an unusual challenge and a tight deadline, performed an overnight modification to some item of construction

TABLE 3–3. Sample of Tractor Shovel Innovations

Basic Innovation: Original tractor shovel	
Major improvements	
Side lift arm linkage	Double-acting hydraulic cylinders
Power steering	Four-wheel drive
Hydraulic bucket control	Torque converter
Fluid transmission coupling	Articulation
Planetary final drive	Power shift transmission
Significant special-purpose accessories	
Lenthened boom arms	Attachment coupler system
Log grapple	Steel-shod tires
Bottom dump bucket	

equipment that solved the problem and saved the day. In fact, however, tractor shovel manufacturers turned out to be the dominant source of commercially successful tractor shovel innovations.

The tractor shovel can be visualized as a four-wheeled, rubber-tired machine with a large, movable scoop mounted at the front end. It is normally used for excavation and other construction tasks as well as for the general handling of bulk materials, ranging from coal to chemicals to soybeans. Householders who live in states with severe winters may have a clear visual image of tractor shovels: They are typically the machines that dig out roads after ordinary trucks have been halted by deep snow.

Tractor shovels are built in many sizes. Today, one can find large tractor shovels with massive, 20 cu-yd scoops working in open-pit mines loading ore into trucks; one can also find small tractor shovels working in warehouses shifting various materials from place to place 1 cu yd at a time. Approximately 41,000 tractor shovels of all sizes were manufactured in the United States in 1980, with an aggregate value of $1.5 billion dollars.[3]

The Sample

The basic tractor shovel was developed in 1939. The sample consists of that basic innovation plus all significant improvements to the tractor shovel commercialized prior to 1970.[4] The major improvements category in Table 3–3 consists of innovations that are installed in virtually all tractor shovels and that are of value to essentially all users. For example, articulation, an innovation that hinges the tractor shovel in the middle and greatly improves steering and traction, is valuable to essentially all users and is now incorporated in almost all tractor shovels. In contrast, the special-purpose accessories listed are innovations that are only of value in some specialized tasks. Thus, the lengthened boom arms are used primarily by those who load high-sided trucks, whereas the log grapples are primarily used by lumber companies in their logging operations. Of course, many other special-purpose accessories exist, ranging from asphalt pavers to snow blowers. The five chosen for study serve relatively large user groups.

TABLE 3–4. Sources of Tractor Shovel Innovations

		Innovation Developed by				
Innovation Type	*% Manufacturer*	*Manufacturer*	*Allied Manufacturer*	*User*	*NA*	*Total*
Basic shovel	100%	1	0	0	0	1
Major improvements	100	10	0	0	0	10
Major accessories	80	2	2	1	0	5
TOTAL	94	13	2	1	0	16

Findings: The Sources of Tractor Shovel Innovation

As can clearly be seen in Table 3–4, almost all of the innovations studied were developed by tractor shovel manufacturers. In two instances these innovating manufacturers were what is known in the trade as allied manufacturers or allied vendors. These are firms that make a specialty of manufacturing attachments for tractor shovels and similar machines. Sometimes they are simply equipment dealers who run a small manufacturing operation on the side. (Tractor shovel manufacturers will often cooperate with such firms because the accessories they build enhance the utility of the basic tractor shovel by tailoring it to various specialized users.)

Only one innovation studied was completely developed by a user: the attachment coupler system, which was developed by a farmer for use on his farm. But users did some innovation work related to some of the other special attachments studied (see the appendix). For example, before steel-shod tires were developed by a manufacturer (they are used to protect tractor shovel tires from cuts), some tractor shovel users were protecting the tires of their machines by wrapping them with heavy steel chains.

Manufacturers as Innovators: Engineering Thermoplastics

Engineering plastics are triumphs of organic chemistry and most were created after World War II. The term engineering plastic simply means a plastic that can be used in demanding engineering applications. Examples of such applications are parts placed under mechanical stress or mechanical shock (e.g., gears or mallet heads) or parts placed in demanding temperatures and/or chemical environments (e.g., parts used in automobile engines). Prior to the advent of engineering plastics, such parts could only have been made of a material like metal or glass. Now they can often be made better and more cheaply from plastic.

All engineering plastics are produced in low volume but with a relatively high selling price when judged against such bulk plastics as polyethylene. In 1976

TABLE 3–5: Engineering Thermoplastics Sample

Innovation (Trade Name)	*U.S. Consumption for Structural Uses*[a] *Millions of Pounds*	*Millions of Dollars*
Polycarbonate (Lexan)	150	143
Acetal homopolymer (Delrin)	20	15
Acetal copolymer (Celcon)	60	47
Polysulfone	12	22
Modified polyphenylene oxide (Noryl)	9	90

[a] 1976 data (James A. Rauch, ed., *The Kline Guide to the Plastics Industry* [Fairfield, N.J.: Charles H. Kline, 1978], 55–58 and Table 3–11).

engineering plastics counted for about 2% by volume of all plastics produced but accounted for about 6% of the total value of all plastics produced.[5]

The Sample

The sample of engineering plastics innovations consists of all commercially successful engineering thermoplastic monomers* introduced to the market after 1955 that achieved sales of at least 10 million lbs annually by 1975. (This definition of commercial success was suggested by plastics manufacturer interviewees.) The five engineering thermoplastics innovations that met these sample selection criteria are identified in Table 3–5.

Findings: The Sources of Engineering Thermoplastics Innovation

As can be seen in Table 3–6, "four and one-half" of the five engineering plastics in the sample were developed by plastics manufacturers. Thus, this very small sample shows a strong manufacturer-as-innovator pattern.

The innovation coded as 50% user developed and 50% manufacturer developed was polycarbonate resin (Lexan), which was developed by General Electric in 1960. GE is both a major producer and a major user of polycarbonate. In the period immediately following commercialization when production capacity was low relative to that of today, GE personnel estimate that as much as 50% of GE polycarbonate production was consumed internally. Currently, GE consumes only a small percentage of annual polycarbonate production. (See the appendix for further details.)

Happily, cases such as GE, where a single firm holds more than one functional role with respect to an innovation, are very rare in our samples. When

*Engineering plastics are characterized as thermoset or thermoplastic resins, with thermoplastics being more commonly used. The two types of plastics are distinguished by the way in which they cure into usable plastic parts. Thermoset plastic forms molecular bonds when molded under high temperature and pressure, and the process is irreversible. Thermoplastics, in contrast, simply "freeze" into a shape on cooling, a process that can be reversed by the simple application of sufficient heat. Monomers are the basic molecular building blocks of plastics.

TABLE 3–6. Sources of Engineering Thermoplastics Innovations

		Innovation Developed by			
Innovation Type	*%Manufacturer*	*Manufacturer*	*User*	*NA*	*Total*
Engineering plastics	90%	4.5	0.5	0	5

dual or multiple roles are held by the same innovating firm or individual, severe coding problems emerge. Often, one cannot determine which role the innovator was "really" motivated by during the development work.

Manufacturers as Innovators: Plastics Additives

Plastics additives are used to modify the properties of a basic polymer in desired ways. An enormous number of additives exist, and they are generally categorized according to the function they perform. Thus, there are coloring agents, flame retardants, fungicides, filling materials, reinforcing materials, and so on. Each of these categories contains a number of materials of varying properties to serve the specified function.

I decided to examine the sources of innovation in two categories of plastics additives: plasticizers and ultraviolet (UV) stabilizers. These two additive types address markets of very different size. (In 1983 more than 600,000 metric tons of plasticizers of all types were sold.[6] In that same year, approximately 2300 metric tons of UV stabilizers of all types were sold.[7] I do not have data on dollar volumes in these two categories: Ultraviolet stabilizer prices are typically somewhat higher than plasticizer prices, however.)

Plasticizers are materials that are incorporated into plastics to improve properties such as workability and flexibility. Without plasticizers, plastics such as polyvinyl chloride (PVC) would be hard and brittle. Ultraviolet stabilizers are added to plastics to protect them from the effect of ultraviolet light such as that present in sunlight. Without such protection, susceptible plastics would quickly discolor, become brittle, or show other undesirable changes.

The Sample

The sample of plasticizer and ultraviolet stabilizer innovations included all commercialized compounds that met four criteria. First, the additive was appropriate for use with at least one of the four largest plastics in commercial use: polyethylene, polyvinyl chloride, polystyrene, and polypropylene. Second, the additive must have been first commercialized after the World War II—a requirement added under the assumption that data on more recent innovations would be of high quality. Third, the additive must have been commercially successful, that is, it had to have been sold on the open market and regarded as a successful product by expert interviewees in the additives

TABLE 3–7. Sample of Plastics Additives Innovations

Plasticizers	
Butyl benzyl phthalate	Tri melitates
2 ethyl hexyl di phenyl phosphate	Tri isopropyl phenyl phosphates
Citroflex type	Epoxidized soybean oils
Di N undecyl phthalate	Long chain aliphatic polyesters
Ultraviolet (UV) stablizers	
2:4 dihydroxy benzophenone	2:4 di t butyl phenyl 3:5 di t butyl phenyl 4 hydroxy benzoates
Ethyl-2-cyano 3:3 diphenylacrylate	Zinc oxide and zinc diethyl dithio carbamates
2 hydroxy 4 dodecyloxy benzophenone	Benzotriozoles
Nickel complexes	
P methoxy benzylidene malonic acid dimethyl esters	

industry. Fourth, the additive was included only if it represented an improvement over previously commercialized additives on a property of importance to users other than cost, for example, decreased toxicity or increased ease of use.

The sample of plasticizers and ultraviolet stabilizers selected as meeting these criteria is identified in Table 3–7.

Findings: The Sources of Plastics Additives Innovation

As can be seen in Table 3–8, more than 90% of the plastics additives innovations studied were developed by firms that manufactured them. (Interestingly, two UV stabilizer innovations coded as manufacturer developed showed a pattern I had not found before: A single manufacturer did not develop the innovation on its own. Instead, an association of manufacturers funded the required R & D work at a private research firm.)

Suppliers as Innovators

Up to this point in the research on the functional sources of innovation, my students and I had conducted six studies in total and had only observed innovation by users and/or manufacturers. But experience gained in these studies was leading me to speculate as to the *cause* of the striking variations in the functional source of innovation that we had been observing.

I will discuss this matter in the next chapter. For present purposes, however, let me just say my speculation was that innovation appeared to be "caused" by potential innovators' relative preinnovation expectations of innovation-related benefit. And, therefore, it seemed to me that innovation in any number of functional loci should exist, given only the proper level and distribution of benefit expectations.

So, Pieter VanderWerf (Ph.D. candidate) and I set out deliberately to find

TABLE 3–8. Sources of Plastics Additives Innovations

		Innovation Developed by				
Additive Type	*%Manufacturer*	*User*	*Manufacturer*	*Supplier*	*NA*	*Total*
Plasticizers	100%	0	5	0	3	8
UV stabilizers	86	1	6	0	1	8
TOTAL	92	1	11	0	4	16

innovation in a third functional locus, suppliers. (Suppliers are firms or individuals whose relationship to an innovation is that of supplying components or materials required in the innovation's manufacture or use.) It seemed reasonable in a rough way that suppliers might develop an innovation that they did not expect to use or sell if that innovation would result in a large increase in demand for something they *did* want to sell. (Thus, gas utilities might develop novel gas appliances and give the designs away to appliance manufacturers, hoping to capture rents from increased gas sales, rather than from making or using the innovative appliance itself.) Based on this logic, we looked for innovation categories that might contain supplier innovation by looking for processes using a great deal of relatively expensive material or components as an input.

Three categories of innovation seemed likely candidates for supplier innovation on the basis of a preliminary inspection, and VanderWerf examined all of them. As the reader will see, all three showed supplier innovation as the hypothesis would predict.

Supplier/Manufacturers as Innovators: Wire Termination Equipment

In a first study, VanderWerf[8] explored the source of innovations in process machines used in electrical wire and cable termination. The idea leading to this focus was simply that many connector designs are unique to particular connector suppliers. Thus, connector suppliers might have a strong incentive to develop machines that would make it more economical for users to apply their products.

Two types of wire termination machines were examined: those that simply cut a wire to length and strip the insulation from its ends; those that cut the wire and attach some sort of connector to one or both ends. Machines used for these purposes range from simple and inexpensive hand tools to complex machines costing as much as $100,000.

Some firms that manufacture wire and cable preparation process machines are equipment manufacturers only. Others manufacture equipment *and* the electrical terminal or connector supply items that the machinery applies to electrical wires. These latter firms have both a manufacturer and a supplier relationship to the process machinery innovations selected for study.

TABLE 3–9. VanderWerf Sample of Innovations in Wire Termination Equipment[a]

Wire type: Single-insulated (hookup) wire

Equipment function: Cutting and stripping

INNOVATIONS
- Automatic cut-and-strip machine
- Linear feed cut-and-strip machine

Equipment function: Nickless stripping

INNOVATIONS
- Thermal stripper
- Die-type hand stripper
- Semiautomatic die-type stripper

Equipment function: Attachment of terminals

INNOVATIONS
- Automatic lead-making machine
- Power crimp bench press
- Strip-fed crimp press

Wire type: Bundled single-insulated wires

Equipment function: Wire stripping

INNOVATIONS
- Rotary stripper

Equipment function: Attachment of terminals

INNOVATIONS
- Stripper-crimper
- Heat shrink-sleeve assembly racks (for soldered connectors)
- Crimp connector assembly machine
- Semiautomatic insulation-displacement terminator
- Automatic insulation-displacement harness maker

Wire type: Ribbon cable

Equipment function: Cutting and stripping

INNOVATIONS
- Automatic ribbon cable cutter
- Automatic ribbon cable cut-and-strip machine

Equipment function: Connector attachment

INNOVATIONS
- Pneumatic ribbon cable press
- Semiautomatic ribbon cable terminator
- Automatic ribbon cable harness maker

Wire type: Flat conductor cable

Equipment function: Attachment of cable terminals

INNOVATIONS
- Crimp stitcher

[a]VanderWerf, "Parts Suppliers as Innovators in Wire Termination Equipment."

The Sample

Different types of wire termination machines are needed to process different types of wire. VanderWerf focused on the machines used to prepare four

TABLE 3–10. Sources of Innovation in Wire Termination Equipment[a]

		Innovation Developed by				
Innovation Type	*%Supplier*	*Supplier[b]*	*User*	*Manufacturer*	*NA*	*TOTAL*
Machines that *do not* attach terminals or connectors	0%	0	1.5[c]	4.5[c]	2	8
Machines that *do* attach terminals or connectors	83	10	0.5[c]	1.5[c]	0	12
TOTAL	56	10	2	6	2	20

[a]VanderWerf, "Parts Suppliers as Innovators in Wire Termination Equipment."
[b]Innovating suppliers were found among connector manufacturers only, not wire manufacturers, except in the case of ribbon cable innovators when both connectors and cable were supplied by innovating firms.
[c]One innovation in this category was developed jointly by a user and a manufacturer, and attributed 50% to each.

types of wire frequently used in electronics equipment: single-strand hookup wire, multiwire round cable, ribbon cable, and flat conductor cable. He then identified an innovation sample that consisted of the first special-purpose equipment used to cut and strip and/or attach connectors to each type of wire as well as the major improvements to this equipment commercialized over the years.[9] This sample is shown in Table 3–9.

Findings

There are firms that specialize in manufacturing wire termination machinery. And, as can be seen from Table 3–10, VanderWerf found that these machine builders were the developers of almost all process machine innovations that did not involve attaching a connector to a wire.

In sharp contrast, almost all of the innovative machines that did attach connectors as part of their function were not developed by these machinery specialists. Rather, they were developed (and manufactured) by the major connector suppliers. We will be able to suggest an explanation for this interesting contrast later when we consider the causes of the variations we have observed in the functional sources of innovation.

Suppliers as Innovators: Process Equipment Utilizing Industrial Gases and Thermoplastics

In a second study VanderWerf[10] examined two additional types of process machinery for evidence of supplier innovation: (1) process machinery that used a large amount of an industrial gas such as oxygen or nitrogen as an input and (2) process machinery that used a large amount of thermoplastic resin as

TABLE 3–11. Sample of Industrial Gas-using Innovations[a]

Basic innovation	Related major improvement
Basic oxygen process	Mixed gas blowing
Detonation gun coating	Electrical sequencing
Nitrogen heat treating	Oxygen probe control
Cryogenic food freezing	Conveyorized freezer
Pyrogenic oxidation	Direct digital control
Argon–oxygen decarburization	Argon–oxygen nitrogen decarburization

[a]VanderWerf, "Explaining the Occurrence of Industrial Process Innovation by Materials Suppliers," Table 5–4, p. 46.

an input. In each of these areas, it seemed reasonable on economic grounds that materials suppliers might develop innovative process machines they did not want to manufacture or use, instead, hoping to capture rents from sales of materials used in the process.

The Samples

For each of the two process areas he would study, VanderWerf chose a sample design consisting of a number of innovation pairs: a major process innovation and a related important improvement. He chose his sample of industrial gas–using process innovations by, first, identifying the 6 industrial gases with the highest U.S. production volume in 1978. Next, he asked experts to identify the single process innovation introduced after World War II that now used the highest volume of each of these 6 industrial gases. (The sample was restricted to post-World War II innovations so that one could collect innovation history data from individuals with a firsthand knowledge of innovation events.) Finally, the single most important improvement to each of these 6 "basic" innovations was also identified by discussions with experts. These were included in the sample as related major improvements. The 12 innovations identified in this way are shown in Table 3–11.

VanderWerf chose a sample of 7 pairs of thermoplastics-using process innovations by a similar procedure. First, the 6 thermoplastics with the greatest U.S. production volumes at the time of the study were identified. Then, the major forming processes used with each as reported by *Modern Plastics* in 1983 were examined, and 7 were found to have been developed post-World War II. These 7 basic innovations were included in the sample. Finally, experts were asked to identify the single most important improvement that had been developed for each innovation up to the present day, and these were included in the sample as related major improvement innovations. These 14 innovations are shown in Table 3–12.

Findings

Careful study showed that supplier innovation did indeed exist. As can be seen in Table 3–13, about one third of both innovation samples had been

TABLE 3–12. Sample of Thermoplastic-using Innovations[a]

Basic innovation	**Related major improvement**
Slush molding	Water bath gelling
Rotational molding	Three-arm RM machine
Direct foam extrusion	Tandem screw extruder
Foam casting	Continuous belt casting
Bead expansion molding	Automatic molding press
Foam closed molding	Cold mold process
Reaction-injection molding	Self-clearing RIM head

[a]VanderWerf, "Explaining the Occurrence of Industrial Process Innovation by Materials Suppliers," Table 5–2, p. 44.

developed by materials suppliers. These firms had only a materials-supply link to these process innovations: They did not themselves either manufacture or use the machinery they had developed.

Additional Evidence on Nonmanufacturer Innovation

The six studies I have just reviewed will be useful when we attempt to understand the causes of variations in the functional sources of innovation, because they have determined the *proportion* of innovations that fall into various loci. Other studies exist, however, to support the *fact* of nonmanufacturer innovation.

Corey[11] provides several case histories of important materials-using innovations developed by materials suppliers. Cases of supplier innovation Corey explores range from the development of vinyl floor tile to the development of fiber-reinforced plastic water piping.

Recently, Shaw[12] examined the role of the user in the development of 34 medical equipment innovations commercialized by British firms. He found that 53% of these were initially developed and proven in use by users. He also found users often aided manufacturers wishing to commercialize their innovations in this field, sometimes helping to design, test, and even market the commercial version.

TABLE 3–13. Sources of Materials-utilizing Process Innovations[a]

		Innovation Developed by				
Material Type	*%Supplier*	*Supplier*	*User*	*Manufacturer*	*Other*	*Total*
Industrial gas	33%	4	5	2	1	12
Thermoplastic	36	5	6	2	1	14
TOTAL	35	9	11	4	2	26

[a]Vanderwerf, "Explaining the Occurrence of Industrial Process Innovation by Materials Suppliers," Tables 8–1, p. 57, and 8–2, p. 58.

Everett M. Rogers and his colleagues explored the diffusion of several innovations used in the public sector and observed that the version initially diffused was often modified by its users. In a study of 10 Dial-a-Ride operations (Dial-a-Ride is a form of shared taxi service), Rice and Rogers[13] report 77 modifications to the original concept in management, technology, or operation (service provided) by adopting users. A similar pattern of user modifications was found by Rogers, Eveland, and Klepper[14] in a study of the diffusion and adoption of GBF/DIME—an information-processing tool developed by the U.S. Census Bureau—among regional and local governments.

Notes

1. Stephen H. Pickens, "Pultrusion—The Accent on the Long Pull," *Plastics Engineering* 31, no. 7 (July 1975): 16–21.

2. Candidate innovations for inclusion in the sample were obtained from a group of experts working for both pultrusion process users and equipment firms. These experts were identified by a two-step process. First, individuals whose names were frequently cited in the technical literature on pultrusion were contacted. Second, the views of these individuals were sought as to the identity of experts in the pultrusion process who had a broad knowledge of the field from its inception to recent times. Individuals so identified (13 in all) were then contacted and asked to identify those innovations that would meet, in their view, the sample selection criteria stated in the text.

The experts polled regarding significant innovations in pultrusion were in substantial agreement with all except the first two innovations listed in Table 3–1. The dissenting experts felt that "true" pultrusion is a continuous process and that this characteristic was not achieved until 1948 with the third innovation listed in Table 3–1. In contrast, the view of other experts and (ultimately) of the author was that these two innovations displayed enough other pultrusion-like characteristics to qualify for inclusion. (See the appendix for further details on these and all other pultrusion innovations studied.) In any case, the decision to include these two innovations did not affect the findings with respect to the sources of innovation in pultrusion.

3. U.S. Department of Commerce, Bureau of the Census, *1977 Census of Manufactures,* vol. 2, *Industry Statistics* (Washington, D.C.: U.S. Government Printing Office, 1980).

4. Candidate innovations for the tractor shovel sample were identified by a two-step method. First, *Engineering News Record,* a widely read trade journal in the civil engineering field, was scanned for the period 1939–74 for mentions of possible candidate innovations. A list of possibilitics developed by this method was then discussed with personnel who had a long history in the tractor shovel industry and who appeared to be both expert and knowledgeable as tractor shovel users or manufacturer personnel. Extensive discussion and numerous deletions from, and additions to, the initial list allowed us to reach a consensus on the innovation sample identified in Table 3–3.

5. James A. Rauch, ed., *The Kline Guide to the Plastics Industry* (Fairfield, N.J.: Charles H. Kline, 1978), 54.

6. "Chemicals & Additives '83: A *Modern Plastics* Special Report," *Modern Plastics* 60, no. 9 (September 1983): 69.

7. Ibid., 60.

8. Pieter VanderWerf, "Parts Suppliers as Innovators in Wire Termination Equip-

ment" (MIT Sloan School of Management Working Paper No. 1289–82) (Cambridge, Mass., March 1982). (Forthcoming in *Research Policy,* entitled "Supplier Innovation in Electronic Wire and Cable Preparation Equipment.")

9. The four wire types studied had annual sales ranging from $35 million to $250 million per year. Major improvements identified by industry experts all were found to provide users with at least a one and two-thirds improvement in labor productivity relative to best preceding practice extant at the time of their commercialization.

10. Pieter A. VanderWerf, "Explaining the Occurrence of Industrial Process Innovation by Materials Suppliers with the Economic Benefits Realizable from Innovation" (Ph.D. diss., Sloan School of Management, MIT, Cambridge, Mass., 1984).

11. E. Raymond Corey, *The Development of Markets for New Materials: A Study of Building New End-Product Markets for Aluminum, Fibrous Glass, and the Plastics* (Boston: Division of Research, Graduate School of Business Administration, Harvard University, 1956).

12. Brian Shaw, "The Role of the Interaction Between the User and the Manufacturer in Medical Equipment Innovation," *R & D Management* 15, no. 4 (October 1985): 283–92.

13. Ronald E. Rice and Everett M. Rogers, "Reinvention in the Innovation Process," *Knowledge: Creation, Diffusion, Utilization* 1, no. 4 (June 1980): 499–514.

14. Everett M. Rogers, J. D. Eveland, and Constance A. Klepper, "The Diffusion and Adoption of GBF/DIME Among Regional and Local Governments" (Working Paper, Institute for Communications Research, Stanford University, Stanford, Calif.; paper presented at the Urban and Regional Information Systems Association, Atlanta, Ga., 31 August 1976).

4

The Functional Source of Innovation as an Economic Phenomenon

We have now seen that striking variations do exist in the functional sources of innovation. Next, we must understand the causes of such variations. Success at this task will allow us to convert the functional source of innovation variable from a "mere" phenomenon into a powerful tool that can be used to explore and manage the innovation process.

I begin this chapter by proposing an economic explanation for variation in the functional source of innovation. Next, I propose two preconditions that must exist in the real world if this hypothesis is to be tenable. A review of the available evidence suggests that the needed preconditions do commonly exist, and this opens the way to a test of the hypothesis in chapter 5.

The Hypothesis

Variation in the functional source of innovation may have many contributing causes. However, we need not necessarily explore all of these to gain a useful ability to predict such variation. I propose that a straightforward economic model will allow us to predict that source "usefully often."

Schumpeter[1] argued that those who succeed at innovating are rewarded by having temporary monopoly control over what they have created. This control, in turn, is the lever that allows innovators to gain an enhanced position in the market and related temporary profits or "economic rents" from their innovations.

Suppose that every firm was aware of the opportunity to develop each of the innovations that I studied in chapters 2 and 3. And suppose every firm analyzed each of these opportunities and said, in effect, "If I respond to this opportunity and develop an innovation, I may expect a rent of x amount over y years." Then my hypothesis, a familiar one to economists, is simply that innovating firms will

TABLE 4–1. Summary of Functional Source of Innovation Data

	Innovation Developed by					
Innovation Type Sampled	*User*	*Manufacturer*	*Supplier*	*Other*	*NA*[a] *(n)*	*Total (n)*
Scientific instruments	77%	23%	0%	0%	17	111
Semiconductor and printed circuit board process	67	21	0	12	6	49
Pultrusion process	90	10	0	0	0	10
Tractor shovel–related	6	94	0	0	0	11
Engineering plastics	10	90	0	0	0	5
Plastics additives	8	92	0	0	4	16
Industrial gas–using	42	17	33	8	0	12
Thermoplastics–using	43	14	36	7	0	14
Wire termination equipment	11	33	56	0	2	20

[a]NA = number of cases for which data item coded in this table is not available. (NA cases exluded from calculations of percentages in table.)

be found among those whose analyses lead them to expect a rent they consider attractive.

This hypothesis will allow us to predict the functional source of innovation only if all or most firms expecting an attractive rent from a given innovation opportunity are of one functional type. There are reasons for supposing that this may often be the case, but let me defer discussion until the end of chapter 5 when we will have seen more data. At the moment, I will simply point out that a summary of the empirical evidence we have collected on the functional source of innovation (Table 4–1) does show that innovations of a specific type are often developed by a single functional type of firm.

Necessary Preconditions

An ability to predict the functional sources of innovation on the basis of firms' preinnovation expectations of rents requires that correlations exist between such expectations and the functional role of innovating firms. More specifically, two conditions must hold:

1. It must be difficult (expensive) for innovators to adopt new functional relationships to their innovations.
2. Innovators must have a poor ability to capture rent by licensing their innovation-related knowledge to others.

In the remainder of this chapter I will explain the necessity for each condition and consider whether each is likely to be commonly present in the real world. Significant empirical data on this matter only exist in the instance of the second condition. However, indirect proof that both conditions at least sometimes hold in the real world is provided in chapter 5 in the form of tests of the hypothesis itself.

Condition 1: Difficulty of Switching Functional Roles

The first condition states that one can only expect to predict the functional source of innovation on the basis of related variations in expected rents if firms do not frequently switch or adopt new functional roles. If role switching were frequent or inexpensively accomplished, innovators might switch to the functional role that offered them the best innovation-related return. And, under such conditions, we would only be able to predict the functional locus of innovation in a weak sense, that is, "the developer of *x* innovation will *become* a user" rather than able to make the stronger statement that "the developer of *x* innovation will be a firm or individual that currently is a member of the user community."

I have no hard data on the general frequency of functional role switching in the U.S. economy. However, I have two reasons for thinking that it is not common. First, as can be seen in the case data contained in the appendix, role switching occurred in only a few instances in the samples we examined. (Typically, in those instances a user firm employee would spin off to form a new equipment manufacturing company based on a user innovation.) Second, the switching of functional roles appears to be often both difficult and expensive.

Consider, as an example, the barriers that face a firm that is a member of an industry characterized by a given functional relationship to an innovation (e.g., an industry using semiconductor process equipment to make semiconductors) and that wishes to join an industry characterized by another functional relationship to that innovation (e.g., the industry manufacturing semiconductor process equipment). These two types of firms are in very different businesses. Each has a great deal of know-how, organizational arrangements, and capital equipment that is quite specialized to build its existing products and to serve its existing customer base. Thus, the semiconductor manufacturer has a sales force that specializes in serving semiconductor buyers. This force would be entirely inappropriate for selling semiconductor process equipment: The customers are different; the sales techniques are different (samples of semiconductor devices can be given out as a selling technique but not samples of semiconductor process equipment); and the specialized knowledge the salesman must have is completely different (a salesman with an electrical engineering background can help customers with problems in selecting and using semiconductor devices whereas a background in solid-state physics would be considerably more appropriate for a salesman trying to sell the semiconductor process equipment used to grow the ultrapure single silicon crystals used in semiconductor device manufacture).

If the sales, organizational, and production infrastructure that a company uses to serve one functional role relationship to a given innovation cannot effectively be used in the service of a different functional relationship, then it follows that a firm wishing to change such relationships must also set up a new infrastructure appropriate to this new role. Further, since the costs of the infrastructures of competitors already having the role relationships the innovator wishes to acquire are typically allocated across many products (e.g., a line of process equipment or a line of semiconductor devices), the would-be new entrant must develop, adopt, or buy a similar line of product to sell if he wishes to be economically competitive. All these requirements, I suggest, can indeed represent significant barriers to the switching of functional roles, although the height of the barriers will, of course, differ from case to case.

Condition 2: Inability to Capture Rent by Licensing Innovation-Related Knowledge

The second condition that must be in place if the hypothesis is to be useful is that innovators must have a poor ability to capture rent by licensing their innovation-related knowledge to others. The reasoning behind this requirement starts with the observation that, barring special situations such as when prizes are awarded for successful innovation, an innovator has only two routes toward capturing rents from an innovation: (1) exploiting the innovation himself while preventing others from doing so; (2) licensing others to use his innovation-related knowledge for a fee.

From the innovator's point of view, it probably does not matter much which of these courses he chooses: Whichever promises the most rent will do. But from our point of view, it matters a great deal. Indeed, the hypothesis must fail if innovators can capture rents efficiently through licensing because it contains the implicit assumption that innovators in different loci have significantly different abilities to capture rent from a given innovation. An ability to license efficiently, however, allows an innovator in any functional locus to (in a sense) tax licensees in different functional loci according to their differing abilities to benefit, thus diminishing this crucial difference.

We can see the problem clearly by means of an extreme example. Suppose that an innovator had a perfect ability to license his innovation-related knowledge without risk or cost. Under these conditions an innovator could license users, manufacturers, or others as he wished. He could set the fee he charged to each innovation beneficiary and each class of beneficiaries so as to maximize his return. Therefore—and this is the crucial point—the functional role that the innovator himself happened to play with regard to the innovation—user, manufacturer, or other—would not influence his expectations of rent because he would be equally able to capture innovation returns from his own company or other companies.

If an innovator is to expect to capture rents from licensing efficiently, he must be able to achieve two things. First, he must have some form of property rights in his innovation-related knowledge that can be used to protect a licensee against those who would want to use the licensed knowledge without paying. Second, he must expect to be able to license others and obtain rent from them at a low risk and cost. In the remainder of this chapter I will focus on the first of these two matters only because an inability to achieve the first makes the second irrelevant. However, when rights to an innovation *can* be protected, then efficiency in licensing becomes very interesting indeed. Since licensing to oneself is costless, any costs or risks incurred in licensing to others will create a preference for in-house use of an innovation and tend to favor those potential innovators who have the largest in-house use for it. (I will return to this matter in chapter 5 when considering the case of engineering plastics innovations.)

Licensing conveys an innovator's rights to innovation-related nonembodied knowledge to another. (Nonembodied knowledge is "pure" knowledge, not incorporated or embodied in physical property such as a product or a process machine.) Currently, only two mechanisms exist in the United States that potentially give an innovator property rights in the nonembodied knowledge he may develop, rights that he may then choose to transfer to another by means of licensing. First, patent law gives an innovator property rights in publicly known information that he has published in his patent. Second, trade secret legislation allows an innovator to license secret knowledge to a user(s) and put the recipient under the legal duty of maintaining the secrecy of that information so that it will not become a free good on the marketplace.

A useful amount of data exists on the real-world effectiveness of these two mechanisms, and I will review it in the remainder of this chapter. If we find that both mechanisms are generally ineffective at protecting innovators' rights to nonembodied knowledge, then the preconditions for licensing that knowledge effectively do not exist and I will have shown that the second condition required for my hypothesis to be useful does often exist in the real world.

Patents and Licensing

A patent grants an inventor the right to exclude others from use of his invention for a limited period. In return for the right to exclude not only those who copy the invention but also those who independently discover the same thing, the inventor must disclose the invention to the public at the time of the patent's issue. This disclosure, contained in the patent itself, must be sufficiently detailed so that those "ordinarily skilled in the art" may copy and utilize the invention after the patent's expiration. While the patent is in force, however, the inventor is given the right to control the use of "his" knowledge.

The real-world value of patent protection to innovators is a much-examined question. A series of studies conducted by several authors over a span of nearly 30 years (1957 to 1984) have asked whether inventors find patents useful for excluding imitators and/or capturing royalty income. The answer uniformly found: The patent grant is not useful for either purpose in most industries. In the next few pages I will briefly review the several studies that came to this conclusion. Then I will explain some of the mechanics of the patent grant and its enforcement that lie at the root of its general ineffectiveness.

Levin et al.[2] conducted a survey of 650 R & D executives in 130 different industries. Several of the questions explored opinions regarding the effectiveness of patents as a means of capturing and protecting the competitive advantages of new and improved production processes and products. The results were that all except respondents from the chemical and pharmaceutical industries judged patents to be "relatively ineffective."[3]

Taylor and Silberston[4] examined the impact of British and foreign patents in a very rich study of 44 British and multinational firms. These firms were selected from five broad classes of industrial activity: chemicals (including pharmaceuticals and petrochemicals), oil refining, electrical engineering (including electronics), mechanical engineering, and man-made fibers.[5]

The upper bound of rents innovators obtain from licensing their patented knowledge can be approximately represented by licensing fees and/or other considerations received, minus patenting and licensing costs incurred by the innovating firm. Taylor and Silberston present such data (Table 4–2) for 30 firms and find these gained little benefit from licensing.

Wilson[6] also studied benefits that corporations reap from licensing their patents. He explored data on royalty payments submitted by some U.S. corporations to the U.S. Securities and Exchange Commission (SEC) in 1971 on Form 10K.[7] Wilson's data for the SIC categories he studied that are most similar to the industrial activity classes examined by Taylor and Silberston are compared in Table 4–3. Here too, corporate returns from licensing appear generally low.[8]

The low returns from licensing patented knowledge found by Taylor and Silberston and by Wilson could be caused by weakness in patent protection or have some other cause. A second type of finding, however, suggests that, whatever the cause, protection afforded by the patent system is in fact generally weak and that innovators in most fields probably could not expect to benefit from licensing their patented knowledge even if they wanted to.

Taylor and Silberston reasoned that, if the patent system does in fact help innovators protect and benefit from their innovations, the presence of patent protection should increase innovators' willingness to invest in R & D. Therefore they asked the companies in their sample: "Approximately what proportion of your R & D in recent years would not have been carried out if you had not been able to patent any resulting discoveries?"[9] The data derived from this question are shown in Table 4–4. Note that 24 of the 32 returns indicate that only 5% or less of recent R & D expenditures would not have been undertaken if patent protection had not been available.[10] This suggests that the patent system is *not* seen by innovators as very effective in general and, by implication, that it is not seen as effective in protecting innovators' rights to knowledge they might wish to license.

A study by Scherer et al.[11] shows a similar result. Only 8 of 37 respondents ("executives responsible for technical change") reported that patents were "very important" to their companies. Fourteen reported that patents had "some importance," and 15 said they were "not very important" to their firms.[12] This result is especially interesting because Scherer selected his sample only from the firms that presumably valued patents most highly—those that held a large number of patents.[13]

In sum, empirical data seem to suggest that the patent grant has little value

TABLE 4–2. Relationship of 1968 Patent Expenditures to 1968 Patent-Related Receipts in Taylor and Silberston's Thirty-Firm Sample

Industry	1 *1968 U.K. license and royalty receipts £ (million)*[a]	2 *1968 U.K. patenting and licensing expenditures £ (million)*[b]	3 *1968 R & D expenditures in U.K. £ (million)*[c]	4 *1968 license receipts as percentage of R & D expenditures plus patenting and licensing expenditures (cols. 1−[2+3])*	5 *1968 license receipts as percentage of 1968 U.K. sales col. 1* *note d*
Chemicals					
Pharmaceuticals	£3.7	NA	£7.1	NA	6.00
Other finished and speciality	0.2	NA	10.1	NA	0.04
Basic	2.4	NA	3.3	NA	1.00
TOTAL CHEMICALS	6.3	0.99	20.5	29	1.10
Mechanical engineering	1.4		7.3	18	0.40
Man-made fibers	0.7	0.37	7.6	9	0.20
Electrical engineering	2.3	0.65	50.5	4	0.30

Source: Adapted from *The Economic Impact of the Patent System: A Study of the British Experience* by C. T. Taylor and Z. A. Silberston. Copyright © 1973 Cambridge University Press. Except as noted in a to d, data in all columns are derived from the same set of companies, NB: Taylor and Silberston have *not* logged patent and R & D expenditure data relative to receipt data on licensing, royalty, and sales.

[a]Taylor and Silberston, Table 8.7, p. 164. (Taylor and Silberst on note that data from oil companies in sample and one large electrical group are excluded from Table 8.7.)

[b]Taylor and Silberston, Table 6.4, p. 109. (Taylor and Silberston note that data from oil companies are excluded from Table 6.4.)

[c]Taylor and Silberston, Table 8.1, col. 2, p. 145. (I have excluded oil company data from basic chemical category to make this data base more compatible with Table 6.4. Taylor and Silberston offer more aggregated R & D expenditure data in Table 6.4, whose magnitudes deviate from those shown in Table 8.1 by 20% to 40%. These discrepancies are unexplained, but my uses of that data are not sensitive to corrections of this magnitude.)

[d]Taylor and Silberston, Table 8.1, col. 4, p. 145.

TABLE 4–3. Wilson and Taylor-Silberston Royalty Data Compared

	Wilson (1971 U.S. data)		Taylor and Silberston (1968 U.K. data)	
Industry	*Percentage of U.S. sales by firms in sample*[a]	*Royalties paid as percentage of firm 1971 sales*[a]	*Royalties paid as percentage of firm 1968 sales*[b]	*Industrial activity*
Chemicals				Chemicals
Industrial	76.4%	0.244%	0.042%	Basic
Drug	72.8	0.745	0.635	Pharmaceuticals
Other	51.4	0.034	0.044	Other finished and speciality
Machinery	40.2	0.051	0.255	Mechanical engineering
Electrical	40.5	0.130	0.182	Electrical engineering

Sources: Adapted from R. W. Wilson, "The Sale of Technology through Licensing" (Ph.D. diss., Yale University, New Haven, Conn., 1975) © 1975 Robert W. Wilson; and from *The Economic Impact of the Patent System: A Study of the British Experience* by C. T. Taylor and Z. A. Silberston. Copyright © 1973 Cambridge University Press.

[a]Wilson, Table 12, p. 169. Note that the data presented here are computed from Wilson's sample of 350 royalty reports, *not* his larger sample comprised of these reports plus estimated data.

[b]Royalty and license-fee expenditures data from Taylor and Silberston, Table 8.7, col. 3, p. 164; sales data from Table 8.1, col. 4, p. 145. Petrochemicals have been removed from the basic chemicals category of Table 8.1 to make this category compatible with the equivalent category of Table 8.7.

TABLE 4-4. Estimated Proportions of R & D Expenditure Dependent on Patent Protection: Twenty-seven Respondent Firms

	Estimate of R & D affected[a]				
Industry	*None or Negligible (number of returns)*	*Very little (less than 5%)*	*Some (5–20%)*	*Substantial (over 20%)*	*Total Returns*
Chemicals					
Finished and speciality	1	2	1	4	8
Basic	1	2	1	0	4
TOTAL CHEMICALS	2	4	2	4	12
Mechanical engineering	7	1	0	2	10
Man-made fibers	1	1	0	0	2
Electrical engineering	7	1	0	0	8
TOTAL	17	7	2	6	32[b]
Percentage of returns	53%	22%	6%	19%	100%

Source: From *The Economic Impact of the Patent System: A Study of the British Experience* by C. T. Taylor and Z. A. Silberston. Copyright © 1973 Cambridge University Press. Reprinted with permission. Table 9.1, p. 197.
[a]Percentages refer to the estimated reduction in annual R & D expenditure in recent years that would have been experienced had patent monopolies not been available.
[b]Some companies made returns for more than one activity.

to innovators in most fields. Are these data congruent with tests of reason? Let us explore.

First, does it make economic sense that firms would take out patents if these do not, on average, yield much economic benefit? The answer is yes, because the cost of applying for patents is also low. The cost of the average patent application prosecuted by a corporation is on the order of $5000 today.[14] Even this small cost is often not very visible to corporate personnel deciding whether to pursue a patent application because it is typically subsumed within the overall cost of operating a corporate patent department.

Second, what do we know about the nature of the patent grant and of the real-world workings of the patent office and the courts? And, is it reasonable in the light of what is known to conclude that the patent grant is likely to offer little benefit to its holder? Consider the following points:

1. A patent, if valid, gives a patentee the right to exclude others from using his invention, but it does *not* give him the right to use it himself if such a use would infringe the patents of others. For example, Fairchild Semiconductor has a patent on the so-called planar process, an important process invention used in the manufacture of integrated circuits. If firm *B* invents and patents an improvement on that process, it may not use its improvement invention without licensing the planar process from Fairchild and in turn that firm may not use the improvement either without licensing it from firm *B*. Thus, in rapidly developing technologies where many patents have been issued and have not yet expired, it is likely that any new patent cannot be exercised without infringing the claims of numerous other extant patents. Given this

eventuality, the benefit of a particular patent to an inventor would very probably be diminished because the patentee might be prevented from using his own invention or might be forced to cross-license competitors holding related patents in order to practice his invention.

2. The patent system places the burden on the patentee of detecting an infringer and suing for redress. Such suits are notoriously long and expensive, and both defendants and plaintiffs tend to avoid them assiduously. For the defendant the best outcome in recompense for all his time and expense is judicial sanction to continue his alleged infringement, whereas the worst outcome would involve the payment of possibly considerable penalties. For the plaintiff the likelihood that a court will hold a patent valid and infringed—as opposed to invalid and/or not infringed—is on the order of one to three.[15] If a patentee has licensees already signed up for a patent at issue, he has a high incentive to avoid litigation: If he loses—and the odds are that he will—he loses payments from all licensees, not just the potential payments from the particular infringer sued.
3. The patent grant covers a particular means of achieving a given end but not the end itself, even if the end and perhaps the market it identifies are also novel. A would-be imitator can invent around a patent if he can invent a means not specified in the original inventor's patent. In the instance of the Polaroid and xerography processes (and a few other notable cases), determined competitors could not in fact invent around the means patented by the inventor. In most instances and most fields, however, inventing around is relatively easy because there are many known means by which one might achieve an effect equivalent to the patented one, given the incentive to do so. Where inventing around is possible, the practical effect is to make the *upper* bound value of an inventor's patent grant equal to the estimated cost to a potential licensee of such inventing around.

Taken in combination, the observations presented above provide a very reasonable explanation for the typical ineffectiveness of the patent grant. However, the data given in Tables 4–2 and 4–3 show clearly that patents are more effective in some fields than in others. This is because the factors mentioned above are more salient in some fields than others, as the following two examples illustrate. First, I will spell out the situation in semiconductors where the patent grant is quite ineffective, and then I will describe the situation in pharmaceuticals where patents have historically been quite effective.

The semiconductor field is a very fast-moving one that contains many unexpired patents with closely related subject matter and claims. The possible consequence—confirmed as actual by corporate patent attorneys for several U.S. semiconductor firms whom I interviewed—is that many patentees are unable to use their own inventions without the likelihood of infringing the patents of others.

Since patents challenged in court are unlikely to be held valid, the result of the high likelihood of infringement accompanying use of one's own patented—or unpatented—technology is not paralysis of the field. Rather, firms in most instances simply ignore the possibility that their activities might be infringing the patents of others. The result is what Taylor and Silberston's interviewees in the electronic components field termed "a jungle" and what one of my interviewees termed a "Mexican standoff."

Firm *A*'s corporate patent department will wait to be notified by attorneys from firm *B* that it is suspected that *A*'s activities are infringing *B*'s patents. Because possibly germane patents and their associated claims are so numerous, it is in practice usually impossible for firm *A*—or firm *B*—to evaluate firm *B*'s claims on their merits. Firm *A* therefore responds—and this is the true defensive value of patents in the industry—by sending firm *B* copies of "a pound or two" of its possible germane patents with the suggestion that, although it is quite sure it is not infringing *B*, its examination shows that *B* is in fact probably infringing *A*. The usual result is cross-licensing, with a modest fee possibly being paid by one side or the other. Who pays, it is important to note, is determined at least as much by the contenders' relative willingness to pay to avoid the expense and bother of a court fight as it is by the merits of the particular case.

Thus in the semiconductor field—except for a very few patent packages that have been litigated, that have been held valid, and that most firms license without protest—the patent grant is worth very little to inventors who obtain it. Indeed, the one value noted to me—defense against the infringement suits of others—suggests that perhaps the true net value of the patent system to firms in the semiconductor industry is negative because it requires all to assume the overhead burden of defensive patenting.

In sharp contrast to the situation pertaining in most other industries and the electronics field in particular, the patent grant often confers significant benefit to innovators in the pharmaceutical field. My discussions with patent attorneys working for pharmaceutical firms brought out two likely reasons for this situation. First, unusually strong patents are obtainable in the chemical field, of which pharmaceuticals is a part. Second, it is often difficult to invent around a pharmaceutical patent.

Pharmaceutical patents can be unusually strong because one may patent an actual molecule found to have useful medical properties *and* its analogs. One need not make each analog claimed but can simply refer to lists of recognized functional equivalents for each component of the molecule at issue. For example, if a molecule has 10 important component parts, one patent application might claim x plus 10 recognized functional equivalents of x for each part. Obviously, by this means an inventor may claim millions of specific molecules without actually having to synthesize more than a few. Furthermore, demonstration that any of the analogs so claimed does not display the medical properties claimed does not invalidate the patent.

Many pharmaceutical patents are difficult to invent around today because the mechanisms by which pharmaceuticals achieve their medical effects are often not well understood. When this is so, potential imitators cannot gain much helpful insight from examining a competitor's patented product. (Interestingly, as biochemists' understanding of the biological basis of the effects achieved by pharmaceuticals improves, one side effect may be to weaken the protection the patent grant affords to inventors of pharmaceuticals.)

Trade Secrets and Licensing

Trade secrets, like patents, can be used as a basis for licensing nonembodied innovation knowledge. A trade secret, also sometimes called know-how, is typically knowledge that can be kept secret even *after* development is completed and commercial exploitation begun. Trade secret legislation allows one who possesses a trade secret to keep the information entirely secret or to make legally binding contracts with others in which the know-how is revealed in exchange for a fee or other consideration and a commitment to keep the information secret. The possessor of a trade secret has an indefinite period of exclusive use of his invention or discovery. He may take legal steps to prevent its use by others *if* he can show that those others have discovered the secret through unfair and dishonest means such as theft or breach of a contract promising to keep it secret.

A legally protectable monopoly of indefinite duration would appear to make trade secrecy a very attractive mechanism for capturing rents from innovation. It is, however, an option only for innovations that can in fact be kept secret: The holder of a trade secret cannot exclude anyone who independently discovers it or who legally acquires the secret by such means as accidental disclosure or reverse engineering. In practice, trade secrets have proven to be effective only with regard to (1) product innovations that incorporate various technological barriers to analysis or (2) process innovations that can be hidden from public view.

There are, in the first instance, certain innovations embodied in products that, while sold in the open market and thus available for detailed inspection by would-be imitators, manage nevertheless to defy analysis for some technological reason and that therefore cannot be reverse engineered. Complex chemical formulations sometimes fall into this category, the classic case being the formula for Coca-Cola. Such barriers to analysis need not be inherent in the product; they can sometimes be added on by design. Thus, some electronic products gain some protection from analysis through use of a packaging method (potting) and packaging materials that cannot easily be removed without destroying the proprietary circuit contained within.[16] Methods for protecting trade secrets embodied in products accessible to competitors need not be foolproof to be effective; they simply have to raise enough of a barrier in a given case to create an unattractive cost-benefit equation for would-be imitators in that case.

In the second instance, process innovations such as novel catalysts or process equipment can be protected effectively as trade secrets, whether or not they could be reverse engineered by a would-be imitator allowed to examine them, simply because they can be exploited commercially while shielded from such examination behind factory walls.

Few empirical data exist on the information protected as trade secrets: There is no central registry for such material analogous to the U.S. Patent Office; even those trade secrets that are licensed to others, the subset of interest to us here, do not usually appear on any public record unless litigated.

Although some examples exist of major rents from nonembodied knowledge being reaped by innovators by means of licensing of trade secrets,[17] I argue that the typical effectiveness of this mechanism is severely limited for two reasons. First, the mechanism is clearly not applicable to product or process innovations that are not commercially exploitable while concealed behind factory walls and that are amenable to reverse engineering if accessible to inspection by imitators—considerations that apply to many industries and many innovations. Second, a trade secret licensor can only gain redress under trade secret legislation if he can document the *specific* illegal act that diffused his innovation to unlicensed parties. A licensor finds such specificity difficult to achieve if he seeks to license nonembodied knowledge to many licensees.

To conclude, it appears likely that both conditions that must exist for my hypothesis to be useful are in fact frequently present in the real world. Therefore, it is appropriate to proceed on to test that hypothesis.

Notes

1. Joseph A. Schumpeter, *Capitalism, Socialism and Democracy,* 3rd ed. (New York: Harper & Row, 1950).

2. Richard C. Levin and Richard R. Nelson, "Survey Research on R & D Appropriability and Technological Opportunity. Part I: Appropriability" (Working Paper, Yale University) (New Haven, Conn., July 1984).

3. Richard C. Levin, "A New Look at the Patent System," *American Economic Review* 76, no. 2 (1 May 1986): 200.

4. C. T. Taylor and Z. A. Silberston, *The Economic Impact of the Patent System: A Study of the British Experience* (Cambridge: Cambridge University Press, 1973).

5. Taylor and Silberston's sample was selected as follows: First, approximately 150 firms selected from a "comprehensive list of U.K. quoted companies" were invited to join the study on the basis of their net assets in 1960. In each of the five classes studied, all companies showing net assets in excess of £10 million in 1960 were selected, and every seventh company of the remainder was selected from a list tabulated in ascending order of net assets in 1960. Finally, "some additions were made to take account of mergers and acquisitions and to include unquoted companies." Eventually, "just over 100" firms responded to the letter of invitation. Sixty-five expressed interest, but "some twenty of these indicated that patents were a very minor aspect of their operations and were firmly believed to have no significance on the business . . . this left 44 firms which agreed to participate in the inquiry" (Taylor and Silberston, *The Economic Impact of the Patent System,* 371).

6. R. W. Wilson, "The Sale of Technology Through Licensing" (Ph.D. diss., Yale University, New Haven, Conn., 1975).

7. In 1971 firms were required to report royalty payments if they were "material" with the precise interpretation of that term being left up to individual firms. Focusing on the *Fortune* listing of the largest manufacturing corporations in 1971, Wilson found that 518 had considered their royalty receipts "material" enough to report to the SEC. Since he was interested only in royalty payments for "technology licenses," he used various means to detect and winnow from the sample firms that reported royalty payments for such things as trademarks, copyrights, and mineral rights (Wilson, "The

Sale of Technology Through Licensing," 152 ff., provides a detailed description of his data collection methods). The end result of this process was a sample of 350 royalty figures for 1971 Wilson felt were largely or entirely payments for "technical agreements," a term that he does not define, but that presumably includes both patent and technical know-how–related payments. The responses of these 350 firms were then aggregated under appropriate "2 and 3 digit SIC codes" (not given) and displayed in tabular form. Wilson used the 350 reports of corporate royalty payments to develop estimates of royalty payments to all members of the industries he studied and then compared these estimates with industry-level data on corporate R & D expenditures collected by the National Science Foundation (NSF). As I find Wilson's estimating procedures inappropriate for my purposes here, I use only the direct company report data he provides.

8. The data reported by Wilson are for royalty payments rather than receipts. But it is likely that the bulk of technical agreements would be between firms in the same industry. If so, it would follow that the low magnitude of royalty payments in the Wilson data implies that royalty receipts would also be found low in the industries sampled.

9. Taylor and Silberston, *The Economic Impact of the Patent System,* 396.

10. Ibid., 30.

11. F. M. Scherer et al., *Patents and the Corporation,* 2nd ed. (Boston: James Galvin and Associates, 1959).

12. Ibid., 117.

13. Scherer sent a questionnaire to a sample of 266 firms shown in P. J. Federico (*Distribution of Patents Issued to Corporations* [*1939–44*], Study No. 3 of the Subcommittee on Patents, Trademarks, and Copyrights [Washington, D.C.: U.S. Government Printing Office, 1957], 19–34) to be corporate assignees for a relatively large number of patents. Sixty-nine of the questionnaires (26%) were completed and returned in time to be included in the study's analysis phase. All but 4 of these respondents held more than 100 patents and collectively they "held approximately 45,500 patents, or about 13.5% of all the unexpired U.S. patents held by domestic corporations at the end of 1956" (Scherer et al., *Patents and the Corporation,* 107).

14. In 1961 the commissioner of patents reported the cost of an average patent application prosecuted by a corporation to be $1000 to $2500, and the cost of a single application prosecuted by an attorney for an individual to be $680 (Hon. David Ladd, Commissioner of Patents, Statement Before the Patents, Trademarks, and Copyrights Subcommittee of the Judiciary Committee, U.S. Senate, September 4, 1962, re: S.2225, quoted in Elmer J. Gorn, *Economic Value of Patents, Practice and Invention Management* [New York: Reinhold, 1964]). My own recent conversations with several corporation patent attorneys yielded an estimate that the average patent application prosecuted by a corporation currently costs on the order of $5000.

15. For references to, and discussion of, several such studies, see Carole Kitti, "Patent Invalidity Studies: A Survey" (National Science Foundation, Division of Policy Research and Analysis) (Washington, D.C., January 1976).

16. Deborah Shapley, "Electronic Industry Takes to 'Potting' Its Products for Market," *Science* 202, no. 4370 (24 November 1978): 848–49.

17. John Lawrence Enos, *Petroleum Progress and Profits: A History of Process Innovation* (Cambridge, Mass.: MIT Press, 1962).

5

Testing the Relationship Between the Functional Source of Innovation and Expected Innovation Rents

In chapter 4 I proposed that it would often be possible to predict the functional sources of innovation on the basis of differences in potential innovators' expectations of innovation-related rents. Now it is time to test this hypothesis. I begin by testing the hypothesis against five samples and find it supported in these instances. Then I draw on the evidence presented to propose that general rules may underlie innovators' expectations of rent.

Five Empirical Tests

In order to test the hypothesis that the functional source of innovation and innovators' expectations of rent are related, we need data on both these factors. Since the work described in chapters 2 and 3 has already provided reliable innovation source data for nine quite diverse innovation types, I found it efficient to create the data base for this test by adding rent expectation data to these same innovation samples where possible. After investigation I found that I could obtain these additional data for five of the nine samples (listed in Table 5–1). These became the basis of the five empirical tests of the hypothesis I have carried out to date.

The inability to use four innovation samples due to data problems is unfortunate. However, the reasons for exclusion do not appear to bias our test of the hypothesis. In the instance of semiconductor process innovations, a single innovation caused multiple product and process impacts whose value I could not isolate and estimate. In the instance of plastics additive innovations and the sample of wire termination equipment innovations, needed data were simply not available from innovating firms or from published sources. Finally, the scientific instrument innovation sample could not be used here because

TABLE 5–1. Summary of Functional Source of Innovation Data

	Innovations Developed by					
Innovation Category	*User*	*Manufacturer*	*Supplier*	*Other*	*NA (n)*	*Total (n)*
Pultrusion process	90%	10%	0%	0%	0	10
Tractor shovel–related	6	94	0	0	0	11
Engineering plastics	10	90	0	0	0	5
Industrial gas–using	42	17	33	8	0	12
Thermoplastics–using	43	14	36	7	0	14

the innovation-related benefits expected by instrument users and instrument manufacturers were different in kind and could not be readily compared. (This is a problem of general interest and I will return to it at the end of the chapter.)

My hypothesis requires that I estimate the rents firms could reasonably expect if they had decided to develop specified innovations. This is not an easy task because expectations of rent are not based on some straightforward calculation. Rather, those who plan to innovate know they must struggle for gain against the sometimes unpredictable actions of competitors, the possible emergence of competing innovations, and other events.

In the battle for innovation-related rents, each firm with an interest in an innovation devises strategies that may help it to minimize innovation costs and maximize its returns from innovation. For example, if Boeing decides to develop a new, more fuel-efficient plane, it will try to lower its innovation costs by shifting some project development expenses to component suppliers and by demanding some advance payments from buyers. Also, it will try to increase its share of the rents generated by the plane by, for example, raising its price to capture some of the fuel savings benefit that users expect to reap. At the same time, of course, suppliers and users are trying to resist these moves and to carry out profit-increasing strategies of their own (e.g., General Electric may attempt to charge more for the fuel-efficient jet engines that it supplies to Boeing).

Given this complex reality, my general strategy for estimating rents that firms might reasonably expect if they were to develop specific types of innovations has been to study several innovation categories in detail and to try to understand the thinking of and options open to potential innovators in these fields.

Innovators capture temporary rents from their successful innovations by first establishing some type of monopoly control over their innovation and then using this control to increase their economic return. A successful innovator's rents may come in the form of cost savings and/or increased prices and/or increased sales obtainable during his period of temporary and partial monopoly. Unfortunately, the available data do not allow us to assign values to these individual components of expected rents or to properly sum and discount them with respect to time. Therefore, in each test that follows I will first summarize information bearing on firms' relative abilities to capture innovation-related rents and then will build a test of reason from this data.

The logic behind my tests of reason will be self-evident, I think, with the possible exception of some of the elements I use to estimate relative abilities to establish monopoly control over an innovation. Let me therefore elaborate a little on the latter before proceeding to a discussion of the tests themselves.

Recall from chapter 4 that we found that innovators typically could not expect to obtain rents from their innovations by licensing them to others because both the patent grant and trade secrecy legislation did not typically allow innovators the type of monopoly control necessary to achieve this. Therefore, innovators must typically benefit by excluding imitators from obtaining rent from their own innovation-related outputs.

The patent grant, trade secret legislation, and response time are the only mechanisms I have observed to date that are exclusively available to innovators and may potentially give an innovator the control over his innovation that he needs to exclude would-be imitators. (This seems a short list, but note that it is not of fixed length. Mechanisms for enhancing an innovator's innovation property rights are social inventions and their number or design has no inherent limit. For example, in the United States one can currently observe the extension of copyright protection to include software writings.)

In chapter 4 we saw that the patent grant generally offers only weak protection to innovators in the context of licensing. I find no apparent reason why it should be any more effective in allowing an innovator to prevent others from imitating his innovation without permission. However, trade secret protection can be much more effective in preventing imitation than it was seen to be in enabling licensing. The difference is that, as a practical matter, licensing of trade secrets requires that they be revealed to licensees, and a secret shared may not stay secret very long. In contrast, a secret kept by an innovator for his own exclusive use need only be known within his factory walls and can often be well protected there.

Finally, an interesting and often effective additional mechanism—response time—became visible in the course of my investigations of innovators' strategies for protecting their innovation-related knowledge. I define response time as the period an imitator requires to bring an imitative product to market or to bring an imitative process to commercial usefulness once he has full and free access to any germane trade secrets or patented knowledge in the possession of the innovator.

Response time exists because many barriers in addition to lack of knowledge must be overcome in order to bring any product or process—even an imitative one—to commercial reality. Engineering tooling must be designed, materials and components ordered, manufacturing plants made ready, marketing plans developed, and so on. During the response-time period an innovator by definition has a monopoly with respect to the innovation and is in a position to capture rent from his innovation-related knowledge.[1]

When an innovator is seeking to protect his innovation from imitators, he may be able to use any or all three mechanisms—patents, trade secrecy, and response time—to prolong his period of exclusive use, his lead time. Whether any or all of these are usable or offer an advantage to a given class of potential innovator is a function of situation-specific factors that will be discussed in the context of the cases that follow.

I begin each case discussion by identifying the types of firm best positioned to capture rents from the category of innovations under study. Next I describe and compare four rent-related elements of the real-world situation facing each type of firm. These are (1) the relative abilities of firms holding different functional relationships to an innovation to establish some monopoly control over it; (2) the nature and amount of innovation-related output generated by innovating and noninnovating firms; (3) the anticipatable cost of innovation; (4) the displacement of existing business that a firm undertaking the innovation studied might expect. Finally, I draw on this information to assess the reasonable preinnovation rent expectations of potential innovators. In this final step I use a very conservative test with respect to my hypothesis. I simply *rank* such rent expectations and count the hypothesis as supported only if the functional source of innovation is populated by firms with the highest preinnovation expectations of rent. (I will return to this point at the end of the chapter.)

Pultrusion Process Machinery: Innovation and Innovation Rents

As we saw in chapter 3, users are the source of all sampled pultrusion process machinery innovations save one. Therefore, the hypothesis I am testing may be stated for this sample as: Innovating users of pultrusion process machinery had higher preinnovation expectations of rent from their innovations than did all firms holding other functional relationships to those same innovations.

There are many classes of firm that have some sort of functional relationship to pultrusion process machinery innovations ranging from inventor to user to manufacturer to distributor. However, it is not necessary to precisely determine the reasonable rent expectations of all of these in order to test the hypothesis. The only noninnovators who can represent a challenge to the hypothesis are those with rent expectations that might conceivably equal or exceed those of the innovating firms. It is therefore efficient to begin this test with a simple inspection of the many extant functional relationships between innovator and innovation and exclude all from further analysis that on the face of it cannot reasonably expect a level of rent from pultrusion process machinery innovations that is anywhere near that of the innovating process users.

In the instance of pultrusion process machinery innovations, I have determined, by means of interviews with industry experts, that the only two functional categories of firm likely to gain significant rent from innovations that improve pultrusion process machinery are process machine users and process machine manufacturers.* I therefore explore the relative ability of these two functional types of firms to capture such rent in what follows.

Relative Ability to Establish Monopoly Control

On the basis of both interview data and reason, it appears to me that innovating users of pultrusion process equipment are better able than innovating manufacturers to establish temporary monopoly control over their innovations. The key source of this difference is the ability of equipment users to hide their innovations for a period of time as trade secrets. This option is not open to manufacturers, who must display their innovations to customers in order to sell them.

Patents and response time were both considered ineffective in this industry by interviewees. Innovators sometimes applied for patents and sometimes received some modest licensing income, but it was understood by all that the patents did not really provide effective protection. Innovators typically would not even attempt to contest infringement of their patents and expected only the naive or exceedingly cautious to honor them. Response time in the instance of pultrusion innovations is only months and is not considered to be of significant value to either user or manufacturer innovators.

Relative Innovation-Related Output

The innovation-related output of the users of pultrusion process equipment innovations is, of course, pultruded product. In general, the effect of the innovations we have studied was to make it possible to extend the pultrusion process to new types or sizes of product. This in turn allowed user firms to create cheaper or better substitutes for products made by other methods or other materials.

*Field investigation showed that the licensing of innovation-related knowledge was (and was considered to be by industry participants) of minor importance in pultrusion processing. Innovators' general inability to appropriate rents from the licensing of knowledge led me to eliminate both independent inventors and suppliers of materials used in pultrusion as potential recipients of significant innovation-related benefit. Independent inventors were eliminated because they only have nonembodied innovation knowledge to sell. Suppliers of materials used in pultrusion were eliminated because pultrusion used 2% or less of the huge amount of polyester resin and fiberglass consumed annually in the manufacture of fiberglass-reinforced plastics during the period of the innovations studied (William G. Lionetta, Jr., "Sources of Innovation Within the Pultrusion Industry" [SM thesis, Sloan School of Management, MIT, Cambridge, Mass., 1977], Table III, p. 41). These innovations needed only commodity plastic resin and fiberglass reinforcement products to implement. Therefore, the suppliers too could not embody innovation-related knowledge in their outputs, and would also be dependent on the licensing of nonembodied innovation knowledge for their innovation-related rents.

Other functional relationships between innovator and pultrusion process innovations—wholesaler and so forth—were eliminated from further consideration after discussion with industry personnel showed innovation benefit potentially accruing to these was clearly much less than that potentially accruing to users and manufacturers.

Pultrusion process machines vary in size, and early machines tended to be smaller and slower than later ones. In 1975 the output of a single average pultrusion machine working a single shift was about 200,000 lb of pultrusions annually. At the 1976 market price of $1.70/lb, this means that a single innovative pultrusion machine could produce $340,000 of novel pultruded product during each year of that machine's service life.

The innovation-related output of a pultrusion machine manufacturer is the incremental hardware on the pultruder that embodies a pultrusion improvement innovation. However, it is more conservative with respect to the hypothesis we are testing to regard innovation-related output as the entire machine embodying such an innovation. Early pultrusion machines tended to be smaller and cheaper than later ones. In 1975 the price of a commercial pultruder of average capacity was $75,000 to $85,000.[2]

Until 1966, all pultrusion process equipment was made by the firms that used it. Then, in 1966 Goldsworthy Engineering began to manufacture a standard line of pultrusion process machines. (Of course, prior to that year there were many manufacturing firms that did produce other types of plastics processing machines and were fully capable of producing pultruders.) By the 1970s each of the largest three user firms had 15 to 20 pultrusion machines each and were making machines for internal use at roughly the same rate as Goldsworthy was making them for the external market. In 1976, pultruder operators reported that about 30 of the 175 pultruders operating in the United States had been built by an equipment builder. The rest were built by users in-house. In that year, pultruders built by users in-house cost less than the commercial equivalent.[3]

Relative Costs of Innovating

The pultrusion process machinery innovations examined were built by both users and manufacturers using general-purpose machine shop equipment that both types of firm had on-site. No organized R & D effort was used to develop these innovations: They consisted of good ideas that, once grasped, could be implemented on the shop floor. Thus it can be assumed that innovation costs would be similar in magnitude for both users and manufacturers of these machines.

Relative Amount of Displaced Sales

No potential innovator in this field had reason to anticipate that pultrusion process innovations would result in a significant displacement of his present sales. In some instances, as can be seen in the innovation cases presented in the appendix, users developed process improvements to reduce their costs of production. In other cases, pultrusions were displacing metals, typically, in high-performance applications. Neither the pultrusion process equipment users nor the equipment manufacturers had any position I am aware of in such displaced products or processes.

Discussion

Now I must speculate. Is it reasonable that even an optimally situated manufacturer of pultrusion machines could expect to gain as much or more rent as innovating users? I do not think so. We have seen that a user has a basis to expect some degree of monopoly control over his innovation, but that a manufacturer does not. Further, I see no relative disadvantage that an innovating user might have with respect to an innovating manufacturer that might offset the user advantage in monopoly control. Users and manufacturers contemplating innovation could expect similar costs, and neither needed to fear displacement of existing business as a result of the type of innovation studied. Further, the possible impact of increasing returns to scale in production of the machines embodying the innovation would be trivial at the levels of production involved here. Even if a single manufacturer produced all of the machines needed by the market (say, 20 per year) his direct per machine costs would only be on the order of 10% less than those of a user producing only a single machine.[4]

In sum, then, the evidence leads me to conclude that the rents that process machine users could reasonably expect prior to an innovation exceed those that a pultrusion process machine manufacturer could expect if he contemplated the same innovation opportunity. Therefore the hypothesis is supported in this instance: Pultrusion process users, the functional type of firm that I judge to have the highest reasonable expectations of innovation-related rent, are also the type of firm that my data show most active in developing pultrusion process equipment innovations.

The Tractor Shovel: Innovation and Innovation Rents

Since, as we saw in chapter 3, manufacturers are the source of almost all tractor shovel innovations, the hypothesis I am testing may be stated for this sample as: Firms that hold the functional relationship of manufacturer to the sampled tractor shovel innovations are also the type of firm best positioned to capture rents from such innovations.

As in the previous study, my first step in testing the hypothesis here was to identify by inspection the few functional types of potential tractor shovel innovators positioned to appropriate significant innovation rents. I found tractor shovel users and tractor shovel manufacturers were clearly the two functional categories of firm most favorably positioned in this regard,* and I will therefore only attempt to rank the relative rent expectations of these two types of firms here.

*My findings here precisely parallel those presented earlier regarding the pultrusion process machinery study. As was the case in that field, innovation benefit was almost never captured from the licensing of nonembodied knowledge. Further, industry participants had no illusions that this could be done, given the general weakness and unenforceability of patents in the field of mechanical invention. As a consequence, I judged that independent inventors were unlikely to be able to

Relative Ability to Establish Monopoly Control

Tractor users and tractor shovel manufacturers apparently are in an equally poor position to establish monopoly control over a tractor shovel innovation they may undertake. Patents did not offer effective protection to innovators in this field. Also, neither user nor manufacturer could expect to protect tractor shovel product innovations as trade secrets. Tractor shovels are used on open construction sites, and any innovations by users and/or manufacturers will be open to the view of would-be imitators. Response times in the instance of tractor shovel innovations are on the order of one year. That is, either a user or a manufacturer of a tractor shovel who had good mechanical skills could imitate a tractor shovel innovation he was able to inspect in a year or less.

Relative Innovation-Related Output

The innovation-related output of tractor shovel manufacturers consists of hardware embodying such innovations. The advantage of the sampled major improvement innovations over previous best practice was such that they were immediately embodied in most or all of the units sold by the innovating firm. Therefore, total tractor shovel sales of the innovating firm in the first post-innovation year is a reasonable indicator of the units of output embodying the innovation in that year.[5]

In the instance of the development of the tractor shovel itself, first-year innovation-related sales were about $250,000 (50 units were sold). All improvement innovations were developed by manufacturers producing at least hundreds of tractor shovels in the year the innovation was commercialized. The major improvement innovations in the sample all added functional capabilities to the tractor shovel at some increase in complexity and cost.

The function of a tractor shovel is to excavate and/or move bulk materials. From a user's point of view, the innovation-related output of a tractor shovel innovation is the increase in productivity that the innovation provides. I estimate the improvement that an innovating user could expect from embodying one of the improvement innovations studied on one tractor shovel is an increase in output of on the order of 20%. This translates into an operating savings of perhaps $1000 annually per machine.[6]

Even the largest users of tractor shovels contemplating an innovation could not expect to incorporate their innovation on more than a very few tractor shovels. Today, with the exception of the U.S. Army and some municipalities, even the major, national account users have a fleet of only 8 to 10 tractor

---

appropriate significant innovation benefit from any tractor shovel innovations they might attempt. The same reasoning suggests that suppliers of components used to implement the innovations were also unlikely to innovate. Components used to achieve an innovative effect when applied to tractor shovels were not themselves novel and were typically available from a number of suppliers as off-the-shelf items.

Industry experts contacted all judged that users and manufacturers ranked highest in the list of fuctional types of firms able to appropriate innovation benefit from tractor shovel innovations.

shovels—and that of mixed models and vintages. Presumably fleets of this size were even more unusual in the period when the innovations we studied were commercialized.

Relative Costs of Innovating

I estimate that user innovation costs would be somewhat higher than those of innovating manufacturers because the equipment and engineering skills needed for innovation are utilized by tractor shovel manufacturers in the course of routine manufacturing and are therefore in place if needed for innovation. In contrast, tractor shovel users have no routine need for such equipment or related engineering skills and may need to acquire them especially for innovation-related tasks.

Relative Amount of Displaced Sales

In the instance of the tractor shovel, users might find that the development of an improvement slightly reduced the value of older tractor shovels and other functionally similar construction equipment that they had in inventory. In contrast, with the exception of Clark Equipment Company (a tractor shovel manufacturer that developed one of the sampled innovations), none of the innovating equipment manufacturers also made construction equipment of similar function, such as bulldozers. Therefore, these manufacturers would anticipate no displaced sales as a result of developing tractor shovel innovations.

Discussion

On the basis of the above discussion, I reason that both users and manufacturers of tractor shovels could only expect to have a monopoly of a year's duration if they chose to develop a tractor shovel innovation. However, it also seems clear that a tractor shovel manufacturer's ability to capture rent on the basis of this period of temporary monopoly is greater than that of any user.

Tractor shovel manufacturers inform me that they generally tend to try to increase sales on the basis of an innovation rather than to increase prices. They feel that they may gain a significant advantage over competitors in a year, possibly selling hundreds of additional tractor shovels due to the presence of the innovation. And, even after imitators enter, innovators' first-to-market reputation may continue to give them a marketplace advantage.[7]

In contrast, it is not clear to me or to users how a year's monopoly on a tractor shovel innovation might significantly benefit innovating users beyond the small operating savings discussed earlier. Because there are many substitute ways to move materials, a given user's unique possession of an innovation would not seem to give him monopoly control over any unique capabilities that would possibly command a high rent from the market.

In this industry, users appeared to have no plausible offsetting advantages that might raise their expectations of rent relative to manufacturers. Indeed, significant returns to scale obtainable from hardware embodying the innovations studied in the tractor shovel market were available to manufacturers but not users.[8]

In sum, then, I conclude that tractor shovel manufacturers are both the functional type of firm likely to have the highest preinnovation expectations of innovation-related rent and also the functional type of firm that in fact did innovate. Thus, I find support for my hypothesis in this sample.

Engineering Plastics: Innovation and Innovation Rents

Since, as we saw in chapter 3, the manufacturers of engineering plastics are the source of almost all innovations sampled in that field, my hypothesis may be stated here as: Manufacturers of the sampled engineering plastics innovations are also the functional type of firm best positioned to capture rents from such innovations.

On the basis of inspection and discussion with industry experts, I concluded that firms with either a user or a manufacturer relationship to engineering plastic resin innovations are most likely to gain significant rents from them.* Therefore, I focus on assessing the relative rents appropriable by each of these two functional groups in this test.

Relative Ability to Establish Monopoly Control

In the instance of engineering plastics, all the innovators studied could and did protect their innovations effectively through patents. This observation fits the general evidence regarding the high effectiveness of patents in the field of chemical inventions[9] and discussion with industry personnel shows that such protection is an important part of the commercial strategy of innovators in the field.

Recall from our earlier discussion in chapter 4 that any innovator could expect similar amounts of rent from a given innovation, given that he had perfect, costlessly enforceable property rights to it. To the extent that the patent protection available in the engineering plastics field has these characteristics, the reasoning applies here and I would expect user and manufacturer innovators to have roughly similar expectations regarding the rents they might obtain from engineering plastics innovations.

However, as was also discussed earlier, in the real world even strong patents do not provide protection that is either perfect or costlessly enforced.

*Although patents offer effective protection in this field, licensing of engineering plastics innovations is unlikely for reasons to be discussed in the text. As a consequence, independent inventors and others without significant innovation-related outputs did not seem to me to be potential appropriators of significant innovation-related rents. Suppliers were eliminated from consideration as potential innovators because the materials used in the manufacture of the plastics studied were commodity chemicals, whose suppliers did not appear well positioned to obtain benefit from innovation-related knowledge through sale of a commodity output.

Indeed, an innovator cannot realistically expect to license his patent rights to others without risk or cost.

Relative Innovation-Related Output

The innovation-related output of an engineering plastics manufacturer is the novel plastic itself. Manufacture of each of the engineering plastics studied is highly concentrated. In my sample the innovator retained a dominant share of the market for his novel product for at least several years. In 1976 Du Pont (the innovator) produced 100% of acetal homopolymer (Delrin), Celanese 100% of acetal copolymer (Celcon), General Electric 100% of modified polyphenylene oxide (Noryl), and Union Carbide 100% of polysulfone. General Electric also produced 75% of all polycarbonate (Lexan) in 1976 and Du Pont produced 57% of all polyamides and 57% of all fluoropolymers.[10]

The innovation-related output of the user of engineering plastics is the products in which these plastics are embodied. I have no data on the volume of engineering plastics consumed by individual users. However, I can report that there are literally thousands of users of each of the innovative engineering plastics studied and that no single user buys more than a small fraction of total production. (The sole exception I identified was in the early days of Lexan production when GE used as much as half of its pilot-plant production internally. This fraction quickly dropped to just a few percent when larger plants were brought on line.)

Relative Costs of Innovating

The costs for the actual innovating firms in this field were either equal to or lower than those that other would-be innovators could reasonably anticipate. All of the innovating firms had substantial ongoing research programs in organic chemistry and thus did not have the significant R & D start-up costs firms without such programs could expect. Some, but not many, users also had such programs in place.

The R & D investment that innovators expend to develop an engineering plastic is orders of magnitude higher than the R & D costs associated with other categories of innovation examined in this chapter. Du Pont's R & D and pilot plant expenditures for Delrin, for example, were $27 million in 1959 dollars.[11] Commercialization is also expensive, because engineering plastics manufacture requires special-purpose plants. Thus, the cost of the first commercial plant for Delrin was $15 million[12]; for Celcon $15 to $20 million[13]; and for Lexan $11 million.[14]

Relative Amount of Displaced Sales

Engineering plastics are intended to be substitutes for other engineering plastics or more traditional engineering materials such as metal and glass.

Potential innovators who currently produce such materials do risk displacement of sales of existing products if they innovate. However, the problem can be minimized or eliminated by proper selection of the properties of the innovative material so that it is not a substitute for the innovator's existing products.

Discussion

As mentioned above, patent protection in this field is good, but there is no reason to expect licensing to be either costless or risk-free for an innovator attempting it. As a consequence, I reason that potential innovators would expect somewhat higher rates of rent for innovation-related output produced by the innovator's own firm. Taken by itself, this fact would not create a higher rent expectation for either user or manufacturer innovators. However, important economies of scale issues tip the balance toward the manufacturer.

Major engineering plastics such as those I studied are manufactured in single-purpose, continuous-flow plants that have significant economies of scale associated with them. As an approximation, processing costs per pound in a plant sized to produce 1 million lb of plastic a year are 10 times those in a plant sized to produce 100 million lb annually.[15] Consider the impact of this factor on a user contemplating developing a new engineering plastic. Since any individual user represents only a small portion of the total demand for an engineering plastic, a user firm considering innovation could not expect to achieve attainable economies of scale by producing for in-house use only. If a user, nonetheless, built a plant sized to fill in-house demand only, it would be in a risky position: Any manufacturer that came up with a functional substitute for the user innovation and produced it in a plant sized to serve the entire market could render the user plant and material uneconomic.

In effect, therefore, both users and manufacturers in this industry are both in a position to control the rights to an innovative engineering plastic that they may develop. But only a manufacturer (or a user who becomes a manufacturer) is in a position to exploit the significant economies of scale associated with engineering plastics manufacture. Given that this is so, a manufacturer that innovates is the only functional type of firm that does not have to incur the cost and risk of licensing this type of innovation to a manufacturer. The consequent saving in licensing-related cost and risk results in a higher expectation of net innovation-related rent for an innovating manufacturer than that which an innovating user might expect.

Process Equipment Utilizing Industrial Gases and Thermoplastics: Innovation and Innovation Rents

The final two tests of the hypothesis were conducted by VanderWerf.[16] These studies were designed precisely to test the hypothesis that rent expectations and the functional sources of innovation are linked and therefore they do not require extra data or analyses to serve my purposes. As a consequence, the review I will present can be relatively brief.

Recall from chapter 3 that by the time VanderWerf began his research, he and I had the hypothesis reported on here in mind. It seemed on the basis of that hypothesis that materials suppliers might be found to innovate under the right conditions. As a consequence, VanderWerf elected to focus on categories of process machinery innovations that used large amounts of material as input *and* where an innovating supplier might hope to have some level of monopoly control over that input material. Under such circumstances it appeared possible that materials suppliers might benefit from and develop innovative process equipment that used their material—even if they did not plan to build or use the innovative equipment themselves. Thus, in the instance of these two studies, we were attempting to predict the functional source of innovation on the basis of assumptions regarding the likely functional source of highest innovation rents. As the reader will see, this experiment worked well.

The Studies

VanderWerf's first sample consisted of innovations in process machines that utilized large amounts of industrial gases as an input, whereas his second sample consisted of innovations in process machines that used large amounts of specific thermoplastics as an input. He began his research by determining, by means of discussions with industry experts, that process machine users, manufacturers, and suppliers of materials processed on the innovative machines all had some reasonable expectations of innovation-related rents. Next, he compared the levels of such benefit that these three functional categories of firm might reasonably expect.

VanderWerf estimated the benefit firms could potentially appropriate from each innovation under study by studying the actual commercial history and innovation-related behavior of users, manufacturers, suppliers, and others. By means of discussions with industry participants, he then estimated the relative appropriable benefits (rent), innovation costs, and the new business fraction (a measure that serves the same function as my displaced sales) each class of would-be innovators could reasonably expect if they had been able to accurately foresee the commercial results actually attained by the various innovations. Possible error in these estimates was compensated for by resolving ambiguity in a direction against the hypothesis under test.

Discussion

For each innovation, VanderWerf ranked four functional categories of innovator (user, manufacturer, supplier, and other) in the order of their expected

TABLE 5–2. Test of Hypothesis; The Source of Innovation Benefit and Innovation Activity Compared[a]

	(A) Industrial gas-using innovations: Predicted probability of innovation			
	Highest	*Second*	*Third*	*Lowest*
Innovator	8	4	0	0
Noninnovator	4	8	12	12
			(chi² $p < .01$)	

	(B) Thermoplastics-using innovations: Predicted probability of innovation			
	Highest	*Second*	*Third*	*Lowest*
Innovator	12	1	1	0
Noninnovator	2	13	13	14
			(chi² $p < .01$)	

[a]VanderWerf, "Explaining the Occurrence of Industrial Process Innovation by Materials Suppliers," 65–66.

level of benefit from that innovation. In the top (bottom) rows of Table 5–2(A) and 5–2(B), VanderWerf positions each firm that actually did (did not) develop each innovation in that expected benefit ranking. As can be clearly seen in Table 5–2, these two samples also strongly support the hypothesis under test.

Conclusions and Discussion

The hypothesis I set out to test was that the functional sources of innovation could be predicted on the basis of potential innovators' expectations of innovation-related rents. We now see that this hypothesis is supported in the instance of the five samples examined (Table 5–3). These test data are encouraging but clearly cannot prove the matter beyond dispute. Nevertheless, I myself find the results encouraging enough to warrant moving ahead to both further research and practical applications.

It would be interesting, for example, to use expectations of rents to predict and empirically explore functional sources of innovation in addition to the user, manufacturer, and supplier sources documented in this book. Thus, wholesale or retail distributors of innovative products, processes, or services should have high expectations of innovation-related rents under some conditions, and they probably will be found to innovate where these pertain.

I speculate that the model will be applicable to industrial products, processes, and services—a very considerable universe. It will also be quite practical in these fields: It requires data on variables that innovators and policymakers in firms may already have at hand.

TABLE 5–3. Is There a Relationship Between the Functional Source of Innovation and Reasonable Expectations of Innovation Rent?

		Number of Innovations found in sources with expected rent rank			
Innovation Type	*Relationship*	*Highest*	*2nd*	*3rd*	*4th*
Pultrusion process	Yes: $p < .02$[a]	9	1		
Tractor shovel–related	Yes: $p < .02$[a]	13	3		
Engineering plastics	Yes: $p < .2$[b] (NS)	4	1		
Industrial gas–using	Yes: $p < .01$[c]	8	4	0	0
Thermoplastics–using	Yes: $p < .01$[c]	12	1	1	0

[a]Chi^2 test. The null hypothesis used in these tests was that innovations would be found equally distributed between the two loci of highest expected rents.
[b]Binomial test (used due to very small sample size). The innovation coded 50% user and 50% manufacturer in Table 3–6 is coded conservatively with respect to the hypothesis here (i.e., coded as 100% *not* in the locus of highest expected rents).
[c]Chi^2 test. Same as note (a) except that VanderWerf was able to rank expected rents reasonably anticipatable by members of four functional loci with respect to each innovation (see Table 5–2).

However, the model will not be practicable in fields where all potential innovators do not measure the rents they expect in commensurable ways. This is so simply because the hypothesis that underlies the model requires that one compare levels of expected rents across functional groups in order to predict the functional source of innovation. And this is only possible in fields where all classes of potential innovators with a significant potential interest in an innovation use commensurable measures. Two examples will illustrate the problem.

In scientific instruments, innovating scientist-users typically work in nonprofit institutions. Usually, their research is supported by governmental agencies who distribute grants on the basis of the expected scientific value of the proposed research rather than on its expected economic value. Therefore, neither users of scientific instruments nor their employers appear to have any reason to measure expected innovation-related benefit in economic terms. Rather, acknowledgement by peers of scientific accomplishment appears to be a major innovation-related incentive for this group.[17] In contrast, scientific instrument manufacturers are profit-making firms and presumably do measure their expected innovation-related benefit in economic terms. Similarly, in the field of consumer goods, consumers are known to evaluate innovations in part in terms of psychological benefits not easily measured in economic terms.[18]

In this chapter I have been able to assess the relationship between innovation and firms' expectations of innovation-related rents only by very careful attention to the details of industry structure and behavior that form such expectations. An important next step in this work, it seems to me, is to move toward generalization by seeking out real-world principles and common strategies—successful and unsuccessful—that underlie innovators' attempts to capture such rents.[19]

The empirical work I have done to date contains clues that might lead to such an ability to generalize. For example, note that the functional source of innovation was shown in the five studies to be populated by firms expecting the highest innovation-related rents (Table 5–3). Since the hypothesis itself proposes only that innovation will be found concentrated among firms who find their expected rents attractive, this result is striking. Perhaps it signals that expectations of rent will often differ significantly between firms holding different functional relationships to a given innovation opportunity. This seems to me to be possible because at the moment I can see two general reasons why the rent expectations of potential innovators could differ significantly as a consequence of the functional relationship they hold to an innovation opportunity.

First, the abilities of firms to protect and benefit from identical innovation-related information can differ as a consequence of functional role. For example, innovating users can often protect process and process machinery innovations as trade secrets better than any other type of innovator. This is so because only users can obtain rent from their innovations while keeping them hidden within their factory walls. Innovators with any other functional relationship to an innovation such as manufacturer or supplier must sell the innovation they develop or persuade others to adopt it before they can benefit. The process of selling or persuading typically involves revealing related secret knowledge to prospective innovation adopters and, as a practical matter, usually destroys the basis for trade secrecy protection.

Also, the risk users face in developing a cost-reducing innovation for their own use may typically be lower than that facing any other functional category of innovator. This is because only users do not have to market such innovations in order to derive rent from them—their own use constitutes a source of such rent. Therefore, the risk to users engaging in an innovation process is that the completed device will not work as intended or will be obsoleted by some other innovation or event. Manufacturers and all other nonusers considering developing that same innovation, on the other hand, face these same risks plus the risk that the innovation will not be accepted in the marketplace.

A second general reason why the rent expectations of potential innovators could differ significantly as a consequence of the functional relationship they hold to an innovation opportunity has to do with industry competitive structure. Firms having user, manufacturer, and supplier relationships to a given innovation often come from different industries. These industries may have different structures, for example, the industry that will manufacture an innovation may be more concentrated than the industry that will use it. Since a firm's innovation-related output is not likely to be determined entirely on the basis of a particular innovation, we can see that expectations of rent are likely to be affected by factors such as preinnovation concentration ratios.

A better general understanding of how firms capture innovation-related rents would have implications beyond innovation. For example, consider the possible impact on current views on when and why firms specialize.

Stigler[20] hypothesizes that "vertical disintegration" will occur as a new industry grows and matures. He reasons that many functions that firms in the new industry require have increasing returns to scale. Initially, the firms in the new industry might perform such functions for themselves because specialization is limited by the extent of the market and because total demand for a particular function might not be great enough to support a specialist firm. But, as the industry grew, demand would increase and eventually it would be reasonable for firms in the industry to spin such functions off to specialist firms that could carry them out on a larger scale and thus more cheaply.

It does seem reasonable to me that Stigler's hypothesis may explain real-world behavior under circumstances in which economies of scale and considerations of production cost are very important. However, different patterns will emerge when innovation-related rents (a factor not included in the Stigler hypothesis) are important—as they often are.

Consider, for example, that users may only be able to obtain rents on innovative process machinery if they build it in-house and protect it as a trade secret. Such rents might be far more significant than any scale-related economies potentially offered by a specialist process equipment manufacturer. Anecdotal evidence exists in support of this idea. Thus: "Most world class German and Japanese manufacturing companies have large, well-staffed, very active machine shops. Much of the success of these companies is a result of the proprietary production processes that are incubated in these shops and therefore unavailable to their competitors."[21]

Notes

1. In principle, if an imitator became aware of an innovator's protected knowledge at the moment he developed it, there would be no response-time protection for the innovator: both innovator and imitator could proceed with commercialization activities in tandem. Response time is an important innovation benefit capture mechanism in reality, however, because would-be imitators seldom become aware of an innovator's knowledge at the moment he develops it. Typically, in fact, an imitator only becomes aware of a promising new product when that product is introduced to the marketplace. Until that point the innovator has been able to protect his product from the eyes of interested competitors inside his factory. After that point, if the product is easily reverse engineered and has no patent protection, only the response-time mechanism can provide the innovator with some quasi-monopoly protection from imitators.

No formal studies yet exist, but the value of response time to innovators can be reasoned to be a function of various situation-specific factors. For example, consider the effect of the length of response time divided by length of the customer purchase decision cycle. A high value of this factor favors the innovator over imitators. Consider one extreme example: a consumer fad item (very short purchase decision time) that sells in high volume for six months only. Assume that the item can be readily imitated but can only be produced economically by mass-production tooling that requires six months to build. Obviously, response time here allows the innovator to monopolize the entire market if he can supply it with his initial tooling. At another extreme is an expensive capital-equipment innovation that customers typically take two years to decide to buy, budget for, and so on—and that competitors can imitate in one year. Obviously, response time in this instance affords an innovator little protection.

2. The source of figures in this section is William G. Lionetta, Jr., "Sources of Innovation Within the Pultrusion Industry" (SM thesis, Sloan School of Management, MIT, Cambridge, Mass., 1977), chap. 2.

3. A homebuilt machine of average capacity (i.e., a machine capable of pultruding product with a cross-section of 6 in. by 7 in.) had a direct cost of at most $60,000 in 1977 (Lionetta, "Sources of Innovation Within the Pultrusion Industry," 43) versus the price of $95,000 charged for an equivalent commercial machine. Presumably the user-built machines were cheaper because the user does not incur some expenses that the machine builder must, such as selling expenses.

4. C. F. Pratten, *Economies of Scale in Manufacturing Industry* (Cambridge: Cambridge University Press, 1971), Table 17–4, p. 173.

5. The primary exception was four-wheel drive. Although it offered advantages to all users, it was a costly feature most advantageous to those operating on difficult terrain and, so, penetrated the market more slowly. The inclusion of four-wheel drive in a tractor shovel added over $1000 in direct cost per unit at the time of the innovation.

6. This estimate of operating savings is derived in the following manner. Standard industry assumptions are that the life of a tractor shovel is five years and that it will operate 2000 hr/yr. Productivity savings primarily involve savings of labor and capital. If we assume an operator was paid $1/hr plus 50% fringe and overhead, which was average at the time of the innovations (Council of Economic Advisers Annual Report 1981, in *Economic Report of the President, Transmitted to the Congress, January 1981, Together with the Annual Report of the Council of Economic Advisers* [Washington, D.C.: U.S. Government Printing Office, 1981], 21–213), an innovation that made a tractor shovel 20% more productive—a reasonable estimate for the major improvement innovations studied—would involve a savings of 20% of an operator's salary (or $600/yr) and a savings of 20% the capital cost of a machine. Since, at the time of the innovations, tractor shovel prices ranged from $5000 to $10,000/unit (or a maximum of $2000 depreciated over five years straight line), total maximum annual savings to the user in capital and labor therefore were $1100/yr/machine/innovation.

7. Glen L. Urban et al., "Market Share Rewards to Pioneering Brands: An Empirical Analysis and Strategic Implications," *Management Science* 32, no. 6 (June 1986): 645–59.

8. Pratten, *Economies of Scale in Manufacturing Industry,* Table 17–4, p. 173.

9. C. T. Taylor and Z. A. Silberston, *The Economic Impact of the Patent System: A Study of the British Experience* (Cambridge: Cambridge University Press, 1973).

10. James A. Rauch, ed., *The Kline Guide to the Plastics Industry* (Fairfield, N.J.: Charles H. Kline, 1978), Table 3–4, p. 59.

11. Herbert Solow, "Delrin: Du Pont's Challenge to Metals," *Fortune* 60, no. 2 (August 1959): 116–19.

12. Ibid.

13. Marshall Sittig, *PolyAcetyl Resins* (Houston, Tex.: Gulf Publishing Company, 1963).

14. National Academy of Sciences, *Applied Science and Technological Progress,* A

report to the Committee on Science and Astronautics, U.S. House of Representatives, GP–67–0399 (Washington, D.C.: U.S. Government Printing Office, June 1967), 37.

15. Wickam Skinner and David C. D. Rogers, *Manufacturing Policy in the Electronics Industry: A Casebook of Major Production Problems,* 3rd ed. (Homewood, Ill.: Richard D. Irwin, 1968).

16. Pieter A. VanderWerf, "Explaining the Occurrence of Industrial Process Innovation by Materials Suppliers with the Economic Benefits Realizable from Innovation" (Ph.D. diss., Sloan School of Management, MIT, Cambridge, Mass., 1984).

17. Robert K. Merton, *The Sociology of Science: Theoretical and Empirical Investigations,* ed. Norman W. Storer (Chicago: University of Chicago Press, 1973).

18. Glen L. Urban and John R. Hauser, *Design and Marketing of New Products* (Englewood Cliffs, N.J.: Prentice-Hall, 1980).

19. Success at this task will allow us to lessen the amount of detail we need and exploit the possibilities of larger data bases such as those recently developed by Keith Pavitt and F. M. Scherer. Pavitt recently examined a data base of over 2000 British innovations to begin to map the origins and flow of technology between sectors of British industry ("Sectoral Patterns of Technical Change: Towards a Taxonomy and a Theory," *Research Policy* 13, no. 6 [December 1984]: 343–73). Scherer recently generated and studied a data base on the interindustry flows of technology based on data from 443 large U.S. industrial companies (*Innovation and Growth: Schumpeterian Perspectives* [Cambridge, Mass.: MIT Press, 1984], 51).

20. George J. Stigler, *The Organization of Industry* (Homewood, Ill.: Richard D. Irwin, 1968).

21. Robert H. Hayes and Steven C. Wheelwright, *Restoring Our Competitive Edge: Competing Through Manufacturing* (New York: Wiley, 1984), 381.

6

Cooperation Between Rivals: The Informal Trading of Technical Know-how

We have seen that variations in the functional sources of innovation can sometimes be explained in terms of potential innovators' relative expectations of economic rents. Since this is so, it becomes worthwhile to think about developing a more general understanding of patterns in the sources of innovation on the basis of a more general understanding of rents and how innovators may seek to maximize them. In this chapter I begin this process by focusing on a mode of cooperative R & D that I call informal know-how trading.

I begin by describing informal know-how trading in general terms. Next, I document how it operates in the field of steel minimill process know-how. Then, I explore how one might explain this form of cooperation in terms of patterns in the rents obtainable from innovation-related knowledge.

Informal Know-how Trading

Know-how is the accumulated practical skill or expertise that allows one to do something smoothly and efficiently. The know-how I focus on here is that held in the minds of a firm's engineers who develop its products and develop and operate its processes. Often, a firm considers a significant portion of such know-how proprietary and protects it as a trade secret.

A firm's staff of engineers is responsible for obtaining or developing the know-how its firm needs. When required know-how is not available in-house, engineers typically cannot find it in publications either: Much is very specialized and not published anywhere. They must either develop it themselves or learn what they need to know by talking to other specialists. Since in-house development can be time consuming and expensive, there can be a high incentive to seek the needed information from professional colleagues. And often, logically enough, engineers in competing firms that make similar products or use similar processes are the people most likely to have that needed information. But are these professional colleagues willing to reveal their proprietary know-how to employees of competing firms? Interestingly, it appears that the answer is quite uniformly yes in at least one industry—and quite probably in many.

The informal proprietary know-how trading behavior I have observed to date can be characterized as an informal trading network that develops between engineers having common professional interests. In general such trading networks appear to be formed and refined as engineers get to know each other at professional conferences and elsewhere. In the course of such contacts, an engineer builds his personal informal list of possibly useful expert contacts by making private judgments as to the areas of expertise and abilities of those he meets. Later, when Engineer *A* encounters a difficult product or process development problem, *A* activates his network by calling Engineer *B*—an appropriately knowledgeable contact who works at a competing (or noncompeting) firm—for advice.

Engineer *B* makes a judgment as to the competitive value of the information *A* is requesting. If the information seems to him vital to his own firm's competitive position, *B* will not provide it. However, if it seems useful but not crucial—and if *A* seems to be a potentially useful and appropriately knowledgeable expert who may be of future value to *B*—then *B* will answer the request as well as he can and/or refer *A* to other experts. *B* may go to considerable lengths to help *A*: for example, *B* may run a special simulation on his firm's computer system for *A*. At the same time, *A* realizes that in asking for, and accepting, *B*'s help, he is incurring an obligation to provide similar help to *B*—or to another referred by *B*—at some future time. No explicit accounting of favors given and received is kept, I find, but the obligation to return a favor seems strongly felt by recipients—" . . . a gift always looks for recompense."[1]

Informal know-how trading can occur between firms that do and do not directly compete. Informal but extensive trading of information with competitive value between direct competitors is perhaps the most interesting case, however, because if we can explain that phenomenon, we can more easily explain trading with less competitive impact. Therefore, I focus much of the ensuing data and discussion on the case of informal know-how trading between direct competitors.

Case Study: Informal Trading of Proprietary Process Know-how Between U.S. Steel Minimill Producers

To date, my data on informal know-how trading between competitors are most complete in the instance of process know-how trading in the U.S. steel minimill industry. I offer this data here as an existence test of the phenomenon and as a means of conveying its flavor.

Minimills, unlike integrated steel plants, do not produce steel from iron

TABLE 6–1. U.S. Steel Minimill Sample

Steel Minimill Firm	*Melt Capacity*[a] *(tons per year, 000)*
Four largest firms	
Chaparral, Midllothian, Texas	1400
Florida Steel, Tampa	1578
North Star, Salt Lake City, Utah	2300
Nucor, Charlotte, North Carolina	2000
Other (randomly selected)	
Bayou Steel, LaPlace, Louisiana	650
Cascade Steel Rolling Mills, McMinnville, Oregon	250
Charter Electric Melting, Chicago, Illinois	130
Kentucky Electric Steel, Ashland	280
Marathon Steel, Tempe, Arizona[b]	185
Raritan River Steel, Perth Amboy, New Jersey	500
Specially selected outlier	
Quanex, Houston, Texas	

[a]*Source:* Nemeth, "Mini-Midi Mills," Table 1, pp. 30–34.
[b]Firm closed in July 1985.

ore. Rather, they begin with steel scrap that they melt in an electric arc furnace. Then, they adjust the chemistry of the molten steel, cast it in continuous casters, and roll it into steel shapes. Modern facilities and relatively low labor, capital, and materials costs have enabled U.S. steel minimill firms to compete extremely effectively against the major integrated U.S. steel producers in recent years. Indeed, they have essentially driven U.S. integrated producers out of the market for many commodity products.

There are approximately 60 steel minimill plants (and approximately 40 producers) in the United States today.[2] The most productive of these have surpassed their Japanese competitors in terms of tons of steel per labor-hour input, and they are regarded as among the world leaders in this process.

Methods

The sample of minimills I studied is a subset of a recent listing of U.S. firms with one or more minimill plants.[3] I selected the four firms with the largest annual molten steel production capacity (melt capacity) from this list and then added six others selected at random from the same list. Later, some interviewees in these firms suggested that I also study Quanex Corporation because it was viewed as an industry outlier in terms of trading behavior; so I also added this firm. All firms included in the study sample are identified in Table 6–1.

Interviews were conducted with plant managers and other managers with direct knowledge of manufacturing and manufacturing process engineering at each firm in the study sample. The questioning, mostly conducted by tele-

TABLE 6–2. Know-how Trading Patterns

Steel Minimill Firm	*In-house Process Development*	*Know-how Trade*
Four largest firms		
Chaparral	Major	Yes
Florida Steel	Minor	Yes
North Star	Minor	Yes
Nucor	Major	Yes
Other		
Bayou Steel	Minor	Yes
Cascade Steel Rolling Mills	Minor	Yes
Charter Electric Melting	Minor	Yes
Kentucky Electric Steel	Minor	Yes
Marathon Steel	Minor	Yes
Raritan River Steel	Minor	Yes
Quanex	Minor	No

phone, was focused by an interview guide and addressed two areas primarily: (1) Has your firm/does your firm develop proprietary know-how that would be of interest to competitors? If so, give concrete examples of process or product improvements you have developed and some estimate of their value. (2) Do you trade proprietary know-how with competitors? With whom? Do you hold anything back? What? Why? Give concrete examples.

The source of major, well-known innovations claimed by interviewees was cross-checked by asking interviewees in several firms, "Which firm developed *x?*" The accuracy of self-reported trading behavior could not be so checked. I, nevertheless, have confidence in the pattern found because interviewees in all but one of the sampled firms provided independent, detailed discussions of very similar trading behavior.

Results

Personnel at all firms except Quanex (selected for study specifically because its behavior differed from the norm) did report routinely trading proprietary process know-how—sometimes with rivals.

Interestingly, reported know-how trading often appeared to go far beyond an arm's-length exchange of data at conferences. Interviewees reported that, sometimes, workers of competing firms were trained (at no charge), firm personnel were sent to competing facilities to help set up unfamiliar equipment, and so on.

Of course, the firms that report informal know-how trading with competitors in Table 6–2 do not trade with every competitor and do not necessarily trade with each other. (The interviewed firms differ widely in technical accomplishment and, as we will see later, a firm will only offer to trade valuable know-how with those who can reciprocate in kind.)

TABLE 6–3. Minimill Costs per Ton (Wire Rod, 1981)[a]

Cost Category	*$ per Ton*	*Percentage of Total*
Labor	$60	21%
Steel scrap	93	33
Energy	45	16
Other operating[b]	65	23
TOTAL OPERATING	$263	
Depreciation	11	4
Interest	7	2
Miscellaneous tax	3	1
TOTAL COSTS[c]	$284	100

[a]*Source:* Barnett and Schorsch, *Steel: Upheaval in a Basic Industry,* Table 4–3, p. 95.
[b]Includes alloying agents, refractories, rolls, and so on.
[c]Excluding any return on equity.

Before turning to consider why the trading of proprietary process know-how occurs in the steel minimill industry, let me examine that behavior in more detail under three headings: (1) Did minimills studied in fact develop/have proprietary process know-how of potential value to rivals? (2) Did firms possessing know-how trade with rivals? (3) Was know-how in fact traded, as opposed to simply revealed without expectation of a return of similarly valuable know-how?

Valuable Know-how?

Since many minimill products are commodities, it is logical that process innovations that save production costs will be of significant value to innovating firms and of significant interest to competitors. Barnett and Schorsch[4] report U.S. minimill 1981 costs to manufacture wire rod (a reasonably representative commodity minimill product) to be as shown in Table 6–3.

On the basis of Table 6–3 data, it seems reasonable that all minimills would have a keen interest in know-how that would reduce their labor and/or energy costs. And, indeed, all those interviewed reported making in-house improvements to methods or equipment in order to reduce these costs. In addition, some reported making process innovations that increased the range of products they could produce.

Nucor and Chaparral conduct major and continuing in-house process development efforts (conducted, interestingly, by their production groups rather than by separate R & D departments). Thus, Nucor is now investing millions in a process to continuously cast thin slabs of steel. If successful, this process will allow minimills to produce wide shapes as well as narrow ones and also will perhaps double the size of the market open to minimill producers—an advance of tremendous value to the industry.

The in-house know-how development efforts of other interviewed minimills are much less ambitious, consisting mainly of relatively small refinements in process equipment and technique. For example, one firm is experimenting with a water-cooled furnace roof that is more horizontal (has less pitch) than that of competitors. (The effect of the flatter furnace roof is expected to be increased clearance and faster furnace loading times, a cost advantage.) Other firms develop modified rollers for their rolling mills that allow them to make better or different steel shapes, and so forth. Although many such process refinements have only a small individual impact on production costs, their collective impact can be large.[5]

In sum, then, most steel minimill firms do appear to develop proprietary know-how that would be of significant value to at least some rivals.

Rivals?

Our next question is: Are steel minimill firms that trade know-how really rivals (direct competitors)? If they are not, of course, the know-how trading behavior we observe becomes more easily explicable: Noncompetitors cannot turn traded proprietary know-how to one's direct disadvantage.*

Many minimills do compete with each other today, although this was not always the case. When minimills began to emerge in the late 1950s to late 1960s, they were usually located in smaller regional markets and were protected by transportation costs from severe competition with other minimills. Today, however, there are many minimill firms and significant competition between neighboring plants. In addition, the production capacity of minimill plants has steadily increased and the larger facilities "define their markets as widely as do integrated [steel mill] facilities."[6]

Some minimill interviewees report that they do trade know-how with personnel from directly competing plants. Others report that they try to avoid direct transfer to rivals, but they are aware that they cannot control indirect transfer. (Since traders cannot control the behavior of those who acquire their information, the noncompeting firms they select to trade with may later transfer that information to competitors.)

Is It Really Trading?

Proprietary know-how is only a subject for trading if free diffusion can be prevented. Therefore, I asked interviewees: Could the proprietary know-how you develop in-house be kept secret if you wanted to do this?

In the instance of know-how embodied in equipment visible in a plant tour, free diffusion was considered hard to prevent. Many people visit minimill plants. Members of steelmaking associations visit by invitation, and association members include competitors. In principle, such visits could be prevented, but the value of doing so is unclear since two other categories of visitors could not be as easily excluded. First, suppliers of process equipment

*Firms that produce identical products may not be rivals for many reasons. For example, firms may be restricted to a regional market by the economics of transport (as with liquified industrial gases or fresh milk products) or by regulation (as with banks and utilities).

often visit plants for reasons ranging from sales to repair to advice. They are expert at detecting equipment modification and are quick to diffuse information around the industry. Second, customers often request plant tours in order to assure themselves of product quality and may notice and/or request information on process changes.

On the other hand, interviewees seem to believe that they can effectively restrict access to know-how if they really want to, and there is evidence for this on a general level. Thus, Nucor and Chaparral both attempt to exert some control over their process innovations, and interviewees at other firms think they have some success. Quanex does not allow plant visits at all and feels it effectively protects its know-how thereby.

Data on this matter are also available at the level of specific innovations, although I have not yet collected them systematically. As an example, however, a firm with a policy of being generally open reported that it nevertheless was able to successfully restrict access to a minor rolling innovation for several years. That firm reported gaining an "extra" $140/ton because it was the only minimill able to roll a particular shape desired by some customers. It apparently only lost control of its innovation when production people explained it to a competitor at a Bar Mill Association meeting.

Interviewees, including top management, were aware of know-how exchange patterns in their industry and emphasized that they were not giving know-how away—they were consciously *trading* information whose value they recognized. Thus, Bayou Steel: "How much is exchanged depends on what the other guy knows—must be reciprocal." Chaparral: "If they don't let us in [to their plant] we won't let them in [to ours]—must be reciprocal." These statements are convincing to me because most interviewees who did engage in information exchange had clearly thought about whom to trade with and why. When asked, they were able to go into considerable detail about the types of firms they did and did not deal with and why dealing with a given firm would or would not involve a valuable two-way exchange of know-how.

Know-how trading in the steel minimill industry was not centrally controlled beyond the provision of general guidelines by top management. Also, no one was explicitly counting up the precise value of what was given or received by a firm, and a simultaneous exchange of valuable information was not insisted on. However, in an informal way, participants appeared to strive to keep a balance in value given and received, without resorting to explicit calculation. On average over many transactions a reasonable balance was probably achieved, although individual errors in judgment are easy to cite. (E.g., in the instance of the minor rolling innovation mentioned above, the innovating firm's sales department was furious when, in their view, engineering "simply gave" the unique process know-how and the associated monopoly rents away.)

Quanex, the Exception

Quanex was the sole exception to the minimill trading norm I found. The firm was not on the list of minimills I used to generate our sample, and I only became aware of it and its outlier status because I routinely asked each firm interviewed if it knew of any firm whose trading behavior differed from its own.

When contacted, Quanex explained its behavior by saying that, first, it did not trade because it felt it had nothing to learn from rival firms (a contention disputed by some interviewees). Second, it said that, while it did produce steel by a minimill-like process, it produced specialty steels and considered its competitors to be other specialty steel producers (e.g., Timkin) and not minimills. And, Quanex reported, it was *not* an outlier with respect to specialty steel producers where, it said, secrecy rather than trading was the norm. (We think this latter point very interesting, but will not pursue it here. If confirmed, it suggests that know-how trading patterns may differ between closely related industries. This in turn opens the way to empirical study of the underlying causes of know-how trading under well-controlled conditions.)

Other Empirical Evidence Regarding Know-how Trading

Is know-how trading unique to the U.S. minimill industry? Or is it a significant form of R & D cooperation in many industries? At the moment, I am aware of only three sources of empirical data on this important matter, and all tend to suggest that informal know-how trading exists in many industries.

First, my students and I have now conducted pilot interviews in several U.S. industries in addition to steel minimills. And, on an anecdotal basis, I can report that we have found informal know-how trading apparently quite common in some industries and essentially absent in others. Thus, self-report by interviewees suggests that trading is widespread among aerospace firms and waferboard manufacturing mills, but rare or absent among powdered metals fabricators and producers of the biological enzyme klenow. Interestingly, however, trading seems a more quasi-covert, secretive activity by engineering staffs in some of these industries than was the case in steel minimills. In minimills, top management was typically aware of trading and approved. This does not seem to be necessarily the case in all industries where significant trading is present.

Second, data in a study by Thomas J. Allen et al.[7] of a sample of Irish, Spanish, and Mexican firms appear consistent with what I am calling informal know-how trading. Allen examined the "most significant change, in either product or process" that had occurred in each of 102 firms during recent years. Interviews were conducted with innovation participants to determine the source of the initial idea for the innovation and important sources of help used in implementation. Coding of the data showed that approximately 23% of the important information in these categories came from some form of personal contact with "apparent competitors" (firms in the same industry).

T. J. Allen elaborates on the behavior observed:

> In a typical scenario, the manager from one of these firms might visit a trade show in another country, and be invited on a plant visit by representa-

tives of a foreign firm. While there he would encounter some new manufacturing technique that he would later introduce into his own firm. In other cases managers approached apparently competing firms in other countries directly and were provided with surprisingly free access to their technology.[8]

Third, Robert C. Allen[9] reports "collective invention" in the nineteenth-century English steel industry—and I think that what he has observed might in fact be an example of informal know-how trading. Allen explored progressive change in two important attributes of iron furnaces during 1850–75 in England's Cleveland district: an increase in the height of furnace chimneys and an increase in the temperature of the blast air pumped into an iron furnace during operation. Both types of technical change resulted in a significant and progressive improvement in the energy efficiency of iron production. Next, he examined technical writings of the time and found that at least some who built new furnaces reaching new chimney heights and/or blast temperatures publicly revealed data on their furnace design and performance in meetings of professional societies and in published material. That is, some firms revealed data of apparent competitive value to both existing and potential rivals, a phenomenon that he called collective invention.

The essential difference between know-how trading and what R.C. Allen calls collective invention is that know-how trading involves an exchange of valuable information between traders that is at the same time kept secret from nontraders. In contrast, collective invention requires that all competitors and potential competitors be given free access to proprietary know-how.[10] Allen finds that this free access requirement presents interpretive difficulties, however.*

As will be seen later when I discuss the causes of know-how trading, the difficulty R. C. Allen notes is not present if the iron manufacturers he examined were actually engaged in know-how trading rather than in collective invention. This seems likely to me. Allen deduced that technical data were made available to all because he observed that much was published and presented to technical societies. Certainly, what was published was public: But know-how with trading value might well have been withheld from publication and/or published only when it had lost proprietary status with the passage of time. Both of these suggested behaviors would be difficult to discern in written records but are in fact part of the trading behavior of present-day firms.

*The interpretive difficulty reported by R. C. Allen: "It is extremely puzzling why firms released design and cost information to potential entrants to the industry. If (as we continue to assume) the industry was competitively organized, it would appear that this action could only rebound to the disadvantage of the firm. To the degree that the information release accelerated technical progress, the price of the product would decline and so would the net income of the firm that released the information" (R. C. Allen, "Collective Invention," 16).

Allen proposes three possible explanations for such behavior (a firm's desire to publicize its accomplishment even at the penalty of lost profit, a firm's inability to keep the know-how secret even if it wished to, speculations regarding special conditions under which a firm might possibly find the open revealing of know-how to be profitable), but the puzzle is not convincingly laid to rest.

An Economic Explanation for Know-how Trading

I propose that it may be possible to explain both the presence and absence of informal trading of proprietary know-how between direct competitors in terms of economic rents accruing to trading firms. I begin by framing the phenomenon in the context of a Prisoner's Dilemma, and then initially explore the plausibility of such a model by reference to the limited amount of real-world information currently available to me.

Know-how Trading as a Prisoner's Dilemma

Let us consider know-how trading between rivals as a two-party Prisoner's Dilemma. It has been shown that the two parties involved in such a Dilemma are likely to achieve the highest private gain over many interactions (moves in the game) if they cooperate.[11] However, each player is continuously tempted to defect from cooperation because he will reap higher returns from a single move if he defects while his partner behaves cooperatively.

Two conditions must hold for a situation to be defined as a Prisoner's Dilemma. The first condition is that the value of the four possible outcomes must be $t > r > p > s$, where t is the payoff to the player who defects while the other cooperates; r is the payoff to both players when both cooperate; p is the payoff to both players when both defect; and, finally, s is the payoff to the player who cooperates when the second player defects. The second condition is that an even chance for each player to exploit and be exploited on successive turns of the game does not result in as profitable an outcome to players as does continuing mutual cooperation (i.e., $2r > t + s$).

Let me begin placing know-how trading in the context of a two-party Prisoner's Dilemma by observing that traded know-how is often possessed by more than one firm prior to a trade. Assume, therefore, that $n - 1$ firms possess a particular "unit" of know-how prior to a given trade. The total rent, R_{total}, that a firm (player) possessing that know-how reaps from it can then be expressed as:

$$R_{\text{total}} = R + \Delta R$$

Here R is the rent that a firm may expect from implementing a unit of know-how if it reveals it to its trading partner and, as a result, n firms possess that know-how. ΔR is the extra increment of rent that the firm can expect to garner if it does *not* trade the unit of proprietary know-how. In that case only $n-1$ firms possess that unit, and the player possessing it therefore gains extra competitive advantage from its use. (In instances when a given unit of know-how is possessed by only one firm prior to a trade and by two posttrade, R will be a duopoly rent and ΔR will be the monopoly rent associated with exclusive possession of the know-how minus the dupoly rent.)

A Base Case

As a base case, assume that in each play of the game two firms both start out with one unit of proprietary know-how unknown to the other. Assume also that each of these two units, although different, has identical R and ΔR associated with it. Then each firm starts with proprietary know-how having a preplay value of $R + \Delta R$.

Because knowledge is the good being traded here, a cooperative trade, r, between the two firms will result in each firm having both units of know-how posttrade, and each having the posttrade rent:

$$R_{total} = 2R$$

That is, posttrade each will have lost that increment of rent, ΔR, that was associated with a more exclusive possession of its own know-how unit, but will have gained the additional rent associated with an additional know-how unit. Similar reasoning allows us to work out the consequences of all four possible outcomes of a single play of the game by the two firms as:

$$t = 2R + \Delta R,\ r = 2R,\ p = R + \Delta R, \text{ and } s = R$$

We therefore find that both the first condition ($t > r > p > s$) and the second condition ($2r > t + s$) required for a situation to be defined as a Prisoner's Dilemma hold if $R > \Delta R$. Therefore, if $R > \Delta R$, a policy of know-how trading will usually pay better in the long run than any other strategy. On the other hand, both conditions fail and continuing defection or no exchange is the best option if $R < \Delta R$.

The simple model just given can obviously be brought into more precise alignment with the real world if we add refinements. But at this point I have only anecdotal data to use in testing, so it is reasonable to defer complexity. Instead, I will attempt to assess the intuitive plausibility of the simple model by reference to real-world examples.

When Proprietary Know-how Offers Little Competitive Advantage

In essence, $R > \Delta R$ holds when the exclusive possession of a know-how unit offers relatively little competitive advantage. This, I suggest, is often the case in the real world. To understand why, it is important to first understand a little more about the actual nature of most (but not all) proprietary know-how.

Know-how may have the ring of something precious and nonreproducible to the nontechnical reader. In fact, most proprietary know-how shares two characteristics: (1) it is not vital to a firm and (2) it can be independently developed by any competent firm needing it, given an appropriate expenditure of time and money. Consider two examples of typical proprietary know-how:

> An engineer at an aerospace firm was having trouble manufacturing a part from a novel composite material with needed precision. He called a professional colleague he knew at a rival firm and asked for advice. As it happens, that competitor had solved the problem by experimenting and developing some process know-how involving mold design and processing temperatures, and the colleague willingly passed along this information.

It was certainly convenient for the firm now facing the difficulty to learn of a solution from the rival, but it was not in any way vital. First, it was possible to struggle along without solving the problem at all. The part was in fact being made but with a high scrap rate and much effort. Second, the engineer assigned to solve this problem was competent and could certainly develop a solution independently, given appropriate time and funds.

> Process engineers at a manufacturer of waferboard (a fabricated wood product somewhat like plywood) were having trouble involving frequent jamming of a production machine with wood being processed. As it happens, competitors had solved this problem by experimenting and developing some process know-how involving the regulation of wood moisture content. When contacted, they passed along what they had learned.

Again, it was convenient for the firm now facing the difficulty to know this solution, but it was not essential or even very important. First, the cost of struggling on without solving the problem at all was not exorbitant: Machine operators could continue to cope simply by stopping the troublesome machine and clearing it as often as necessary. Second, a competent engineer assigned to solve this problem could certainly solve it independently.

When proprietary know-how does have the attributes just described, we can perhaps intuitively see the plausibility of the model's prediction that rival firms will find it profitable to engage in know-how trading. Conceptually, the consequences of noncooperation in know-how sharing under such conditions are similar to those of a policy of not cooperating in sharing spare parts with rivals who use an identical process machine. An industrywide policy of noncooperation between rivals with respect to spares would under most circumstances not permanently deprive any firm of needed spares nor otherwise significantly affect the competitive position of rival firms in the industry. It would simply result in increased downtime and/or spares-stocking costs for all—a net loss for all relative to the consequences of a policy of cooperation.

When Proprietary Know-how Offers Significant Competitive Advantage

Sometimes, the competitive value of a unit of know-how is large, and $R < \Delta R$. According to the model, one would then expect that informal know-how trading would *not* occur. I can illustrate this possibility with an interesting example that appears to show know-how trading behavior shifting as the value of a given type of know-how shifts over time.

Aerospace engineer interviewees have informed my students and me that they freely exchange most know-how under normal conditions. But, when a competition for an important government contract is in the offing, the situation changes, and trading of information between rivals that might affect who wins the contract stops. Later, after the contract has been awarded, the same know-how that was recently closely guarded will apparently again be traded freely.

The reported behavior seems reasonable. Much aerospace know-how has the characteristics discussed earlier: It is not critical and under normal conditions it can be independently reproduced by competent engineers if need be. Therefore, it is likely that $R > \Delta R$ for such know-how, and that know-how trading would pay according to the model. But when a competition for an important government contract is near, conditions are not normal. Often, there will not be enough time to produce needed know-how independently, and therefore the ΔR value of a given piece of competition-related know-how could increase temporarily. If the increase reached the point where $R < \Delta R$, it is reasonable according to the model that know-how trading temporarily stop—the behavior in fact reported by interviewees. And, of course, after the contract is awarded, it is reasonable that the ΔR value of competition-related know-how will drop and trading resume—as interviewees report that it does.

In the example just given the know-how at issue could have been independently redeveloped by anyone who wanted it. But the know-how nonetheless yielded competitive advantage to its possessor because the time needed for independent redevelopment was simply not available. Sometimes, however, know-how that can yield a major competitive advantage cannot be routinely reinvented. (It may, for example, be the result of unusual insight and/or major research efforts.) Then, $R < \Delta R$ for years, and trading of that know-how may never be in the best interests of the firm possessing it.

When Proprietary Know-how Offers No Competitive Advantage

Unique possession of proprietary know-how offers essentially no competitive advantage to a firm with respect to nonrivals. Therefore I would expect know-how trading to be to the advantage of firms in such a situation (assuming that the traded information does not leak from nonrivals to rivals) and would predict it to occur. Anecdotal evidence available to this point supports this prediction, but it is certainly only of illustrative value. For example, on the basis of interviews, I find that electric and gas utilities (which serve different regions and are therefore not rivals) do appear to share know-how extensively.

When Diffusing Proprietary Know-how Has Competitive Value

In at least some real-world conditions it appears that competition is enhanced by wide sharing of some know-how. As an example consider the establishment of uniform standards in a product category that can sometimes enlarge markets and benefit all participating manufacturers (e.g., standards set for computer networks and compact audio disks). The establishment of such standards requires some sharing of know-how by participating firms. As a second example consider the sharing of proprietary information on safety hazards between rivals (e.g., on the hazards of dioxin among rivals in the chemical industry).

Informal Know-how Trading in Context

Informal technology trading can usefully be compared with and contrasted to two other forms of R & D exchange between firms: (1) agreements to perform

R & D cooperatively and (2) agreements to license or sell proprietary technical knowledge. As we will see, informal know-how trading can usefully be seen as an inexpensive, flexible form of cross-licensing. Under appropriate conditions, it appears to function better than either of these better-known alternatives.

Agreements to trade or license know-how involve firms in less uncertainty than do agreements to perform R & D cooperatively. This is because the former deals with *existing* knowledge of known value that can be exchanged quickly and certainly. In contrast, agreements to perform R & D offer *future* know-how conditioned by important uncertainties as to its value and the likelihood that it will be delivered at all. The value of the know-how contracted for is uncertain because R & D outcomes cannot be predicted with certainty. The delivery of the results of cooperative R & D projects to sponsoring firms is somewhat uncertain because such results are best transferred back to the sponsoring firms in the minds of employees participating in the cooperative research. Given the U.S. tradition of frequent job changes, participants run significant risk of losing the benefits of their investment by losing the employee(s) they assigned to the project.

Informal know-how trading such as that reported here has a lower transaction cost than more formal agreements to license or sell similar information. Transaction costs in informal know-how trading systems are low because decisions to trade or not trade proprietary know-how are made by individual, knowledgeable engineers. No elaborate evaluations of relative benefit or seeking of approvals from firm bureaucracies are involved. Although informal, each engineer's assessment of the relative likely value of the trades he elects to make may be quite accurate: An information seeker can tell on the basis of his first interaction whether the expert advice he is given is of good quality—because he will immediately seek to apply it. An information provider can test the level of the inquirer's expertise and future value as a source of information by the nature and subtlety of the questions asked. Also, although a particular informal judgment of the value of a trade may be quite incorrect, many small transactions are typically made. Therefore, the net value of proprietary process know-how given and received will probably not be strongly biased for or against any participating firm.

In general, informal know-how exchange between rival and noncompeting firms is the most effective form of cooperative R & D when (1) the needed know-how exists in the hands of some member of the trading network, when (2) the know-how is proprietary only by virtue of its secrecy, and when (3) the value of a particular traded module is too small to justify an explicit negotiated agreement to sell, license, or exchange. Taken together, the second and third conditions have the effect of insuring that the know-how recipient will be free to use the information he obtains without fear of legal intervention by the donor firm. Since much technical knowledge key to progress consists of small, incremental advances, the universe bounded by these three conditions is likely to be a substantial one.

Formal know-how sale or licensing is likely to be preferred when the know-how in question (1) already exists and (2) is of considerable value relative to the costs of a formal transaction. Experts in the oil and chemical industries report that they may engage in formal licensing and sale rather than informal exchange precisely because the value of the know-how in question is typically very high.

Agreements to perform cooperative R & D must be the form of cooperation of choice when (1) the needed information does not exist within any firm willing to trade, license, or sell, and when (2) individual firms do not find it worthwhile to develop modules of the needed know-how independently. This would occur when know-how modules have no profitable applications as modules. Perhaps this is often the case, but I am not sure. Perhaps most "new" know-how in fact consists largely of existing modules of know-how developed for other purposes.

Discussion

Up to this point, I have discussed informal know-how trading as a firm-level phenomenon involving the trading of innovation-related know-how between technical personnel. But the model of such trading that I have presented here contains no inherent restriction as to the nature of know-how traded or as to the nature of the trading parties. Perhaps, therefore, the phenomenon exists and makes sense for individuals and other types of organizations and for other types of know-how as well. A certain answer must await appropriate research, but there are intriguing suggestions that informal know-how trading may be quite general. For example, Collins[12] has shown that scientists employed by nonprofit laboratories (university and governmental) selectively revealed data to colleagues interested in know-how related to the TEA laser. He noted that individuals and laboratories made conscious and careful discriminations as to what know-how would be revealed to what recipient, and he noted also that "nearly every laboratory expressed a preference for giving information only to those who had something to return."[13]

In arenas where know-how trading is applicable, what is its significance? An answer to this question also awaits further research. However, it seems to me possible that it may be an important phenomenon in some arenas. For example, Mansfield[14] recently found a general pattern of rapid transfer of proprietary industrial information from the firms that generated it to others, and he suggested that this might be caused by uncompensated "leakage" of such information to the detriment of the originating firms. But is it perhaps, instead, an indicator of massive know-how trading? If the observed information transfer is indeed simple leakage without compensation to the information generator, then, as Mansfield suggests, innovators face very serious appropriability problems. If, on the other hand, the rapid transfer observed is the result of information trading such as that present in the steel industry, then we may be observing a phenomenon that actually increases firms' ability to appropriate benefit from technical know-how.

Whatever the generality of know-how trading turns out to be, I am sure that further study will also show it can be quite an elaborate phenomenon. Thus, we will surely find know-how trading strategies more complex than those envisioned in a simple, two-party Prisoner's Dilemma, and we may find multiple layers of trading incentives and strategies active in a single trading entity as well.

One obvious form of know-how trading strategy builds on the observation that many firms often have a unit of know-how that a trader needs—and some of these potential trading partners may be direct rivals and some not. I have focused on trading between rivals here simply because it is the costliest form of trade and thus potentially the hardest to explain as economically rational behavior. However, in the real world it is likely that firms would prefer to trade know-how with nonrivals because traded information may then have less or no competitive cost.

Second, consider that firms can form coalitions with respect to know-how trading and restrict that activity to only a subset of firms in their industry. This can be profitable under some conditions. For example, the members of such a club may collectively face a more elastic demand than is faced by the industry as a whole and therefore may gain greater returns from (cost-reducing) innovations. Thus, U.S. or Japanese semiconductor producers may decide it is to their advantage to trade know-how with other domestic firms but not with foreign firms—or vice versa.

Third, consider that strategies may exist that are possible because the substance of know-how trades is knowledge. For example, firms may find a strategy of relatively *rapid* know-how trading may pay dividends. Such a strategy is based on the assumption that a firm receiving know-how in trade does not care who originally developed it: The recipient only cares that it has value to him. Since only novel know-how is valuable to a recipient (there is no value in getting the same information twice), a strategy of rapid know-how trading might allow a firm to exchange its own know-how *and* the know-how developed by others (obtained from earlier trades) to firms that still find that know-how novel, a trading advantage.

As an example of multiple levels of trader existing within a given trading entity, consider that trading between firms such as that I have documented here must also involve a different level of trader—the individuals who actually conduct the trades. It is clear that the benefits to individuals actually engaged in the trading *may* differ from those of the firms that employ them. (But they do not necessarily differ. Consider that an engineer's motive in trading may be in part to improve his potential marketability to competing firms. In this case, a strategy of being helpful to colleagues employed by competitors without hurting the interests of one's present firm by revealing vital proprietary secrets might be optimal for the individual trader as well as for the firm since no one wants to hire someone with a penchant for betrayal.) Research may show that the benefits expected by the different active interests in a trading entity are correlated in important arenas. When this is the case, simple models such as the one presented here may provide us with a practical ability to predict the role of know-how trading in the distributed innovation process.

Notes

1. *The Havamal, with Selections from Other Poems in the Edda,* as quoted by Marcel Mauss in *The Gift: Forms and Functions of Exchange in Archaic Societies,* trans. Ian Cunnison (Glencoe, Ill.: Free Press, 1954), xiv. Mauss has studied patterns of gift giving in a number of cultures and finds the practice typically associated with strong obligations for recompense to be provided by the recipient of a "gift."

2. Early minimills were relatively small (50,000–150,000 tons per year capacity) and produced primarily commodity products such as the reinforcing bar used in the construction industry. Today, however, some individual plants approach 1 million tons annual capacity and many are reaching far beyond commodity products into forging quality, alloy steel, stainless steel, and "nearly any steel grade capable of being melted in an electric furnace" (Edward L. Nemeth, "Mini-Midi Mills—U.S., Canada, and Mexico," *Iron and Steel Engineer* 61, no. 6 [June 1984], 27).

3. Nemeth, "Mini-Midi Mills."

4. Donald F. Barnett and Louis Schorsch, *Steel: Upheaval in a Basic Industry* (Cambridge, Mass.: Ballinger, 1983).

5. Samuel Hollander, *The Sources of Increased Efficiency: A Study of Du Pont Rayon Plants* (Cambridge, Mass.: MIT Press, 1965).

6. Barnett and Schorsch, *Steel: Upheaval in a Basic Industry,* 85.

7. Thomas J. Allen, Diane B. Hyman, and David L. Pinckney, "Transferring Technology to the Small Manufacturing Firm: A Study of Technology Transfer in Three Countries," *Research Policy* 12, no. 4 (August 1983): 199–211.

8. Ibid., 202.

9. Robert C. Allen, "Collective Invention," *Journal of Economic Behavior and Organization* 4, no. 1 (March 1983): 1–24.

10. Ibid., 2.

11. Robert Axelrod, *The Evolution of Cooperation* (New York: Basic Books, 1984).

12. H. M. Collins, "Tacit Knowledge and Scientific Networks," in *Science in Context: Readings in the Sociology of Science,* ed. Barry Barnes and David Edge (Cambridge, Mass.: MIT Press, 1982), 44–64.

13. Ibid., 59.

14. Edwin Mansfield, "How Rapidly Does New Industrial Technology Leak Out?" *Journal of Industrial Economics* 34, no. 2 (December 1985): 217–23.

7

Shifting the Functional Source of Innovation

We have now found that differences in the functional source of innovation may be caused by potential innovators' differing expectations of innovation-related rents. If this is so, we may be able to *shift* the likely source of innovation by changing the distribution of these expectations of rent. Further, by understanding how expected innovation rent is distributed, we may be able to *predict* the likely source of innovation. And if we can do these two fundamental things, we are well on the way to learning how to manage an innovation process that is actually or potentially distributed across different functional loci.

In this chapter, I report on a natural test of the possibility of shifting the functional source of innovation. In chapter 8, I report on a test of the possibility of predicting the sources of innovation.

Nature of the Test

The test I present here deals with a variable that is under the control of firm managers: product design. It starts with the observation that product manufacturers can design products that are easy (inexpensive) or difficult (expensive) for users to modify. An easily modified product will lower a user's expected costs for innovations that require such modification. In contrast, of course, a product that is expensive for users to modify should raise users' expected costs. Such differences in expected costs should in turn affect users' expectations of innovation-related rents and cause differences in the amount of innovation activity involving product modification.

We should be able to observe this hypothesized shift in the functional source of innovation if we can contrast two products—one difficult for users to modify and one easy for users to modify—that are otherwise identical in function and market. Stan Finkelstein and I explored such a situation in the clinical chemistry-analyzer market.[1]

Automated clinical chemistry analyzers are used in clinical laboratories to determine the level of a chemical such as glucose in blood. These automated machines execute a clinical chemistry test method by first combining a sample of blood serum with one or more reagents and then allowing the chemical reaction(s) thus initiated to take place under prescribed conditions of time and temperature. Substances that would interfere with the test measurement are removed (by precipitation, dialysis, or other means), the test measurement is made (through techniques such as colorimetry, fluorometry, etc.), and, finally, the test result is recorded. Automated clinical chemistry analysis has been widely adopted in the United States.[2]

This study focuses on the three brands of automated clinical chemistry analyzer equipment that were most frequently present in hospital clinical chemistry laboratories at the time of our study: Technicon, Du Pont, and Abbott.[3] Each of these analyzers was designed by its manufacturer to perform a number of common tests. Users who wish to use them to develop and perform other tests must use different chemicals and/or modify the analyzers themselves to achieve their goals.

According to clinical chemistry autoanalyzer users and manufacturers, Technicon and Abbott analyzers were much easier to adapt to new test development than were Du Pont analyzers. The cause of this difference lay in the design of each product's reagent handling system. Let me briefly describe that aspect of each brand's design to make the matter clear.[4]

Technicon automated clinical chemistry analyzer models are based on a principle called continuous flow analysis, and they function much like miniature, continuous-process chemical plants. They consist of functional modules—for example, pump modules, dialyzer modules, and so on—interconnected by plastic tubing. Reagent is placed in bulk reservoirs and metered into the system as needed.

Abbott Laboratories automated clinical chemistry analyzers meter the amount of reagent(s) needed for a particular test from bulk reservoirs into transparent, disposable, open-topped plastic cups called cuvettes. Samples of patient serum are also metered into these cuvettes and the desired test reaction proceeds.

Finally, the Du Pont aca clinical chemistry analyzer uses reagents supplied in single-use, disposable, factory-sealed test packs. These are quite complex. Each contains a plastic pouch divided internally into several sealed compartments that contain reagent quantities needed for a single execution of a particular test. The pouch itself is sealed to a plastic header that contains a serum inlet valve and, for tests that require it, a built-in chromatographic column. All chemical reactions required for a test occur inside the disposable test pack; the pack itself is never opened during its transit through the analyzer equipment.

On the basis of the above capsule descriptions, the reader may find it reasonable that users could experiment with novel test methods and equipment configurations by using Technicon or Abbott Laboratories equipment at a lower cost than could be done by using Du Pont equipment. Technicon modules may be purchased and connected up in a novel configuration. In both Technicon and Abbott equipment, desired novel reagents can be mixed up in bulk, placed in the machine's reservoirs, and the *machine* will meter out the proper amount of reagent(s) and serum needed for each test.

Setting up the same novel method on Du Pont equipment, on the other hand, requires buying empty test packs from Du Pont (empty packs without chromatographic columns are for sale—these have a standard use in machine calibration). The experimenter would then inject precisely measured amounts of reagent into selected compartments of each pack and reseal each compartment. If 1000 tests were required for an experiment, experimenters would have to perform these operations on 1000 packs. This would clearly be a great effort, and the end result would be the accomplishment of a reagent proportioning task Technicon and Abbott Laboratories equipment does automatically.

The Test

Users employ automated clinical chemistry analyzers in research that (1) do and (2) do not utilize tests available commercially from the analyzer manufacturer. In the latter case, the needed test is developed by an equipment user. Since (as we saw) Du Pont equipment is relatively difficult for innovating users to modify, we hypothesize that research projects using Du Pont analyzers will involve a lower proportion of user-developed tests than will research using either Technicon or Abbott analyzers. That is, we hypothesize that one particular type of user innovation activity only, that involving modifications to equipment or test reagents, will be shifted away from Du Pont analyzers owing to their design.

Stan Finkelstein and I tested this hypothesis through a search of the medical research literature. If the hypothesis was correct, we would expect that the ratio of research reports involving commercial versus user-developed test methods should be significantly higher when Du Pont analyzers are used in the research than when Technicon or Abbott Laboratories analyzers are used.

First, we searched the medical literature through MEDLINE, a computerized index of approximately 3000 biomedical journals, to identify all research articles that reported using a Technicon, Abbott Laboratories, or Du Pont clinical chemistry autoanalyzer.[5] Next, we read the identified articles and coded those that did in fact use one of these analyzer brands shown in Table 7–1.

As can be seen in Table 7–1, the results of our test support the hypothesis. Thus, it does appear that product manufacturers can influence the amount of user-innovation activity related to their products by raising or lowering the cost of such activity, in this instance, through product design.

TABLE 7–1. Frequency of User Research Articles Involving Commercially Supplied Versus User-Developed Chemical Methods as a Function of Manufacturer of Analyzer Used

Number of articles found reporting research by user personnel only that involved:[a]		
(A) Manufacturer-commercialized chemistries[b]	*(B) Researcher-developed chemistries*	*Performed on automated clinical chemistry analyzers manufactured by*
20	22	Technicon
6	0	Du Pont
5	6	Abbott Laboratories
	Null hypothesis that (B)-type research as likely on Du Pont analyzers as others is rejected: Du Pont vs. Technicon $p = .02$; Du Pont vs. Abbott Labs $p = .04$ (Fisher exact)	

[a]Since our goal is to determine *user* ability to, and interest in, modifying manufacturer-supplied chemistries for the analyzer brands listed, papers written by manufacturer personnel only or written jointly by manufacturer and user personnel are excluded. One paper thus excluded was written jointly by a Du Pont and a user research team and reported a researcher-developed chemistry for the aca (Gopal S. Rautela and Raymond J. Liedtke, "Automated Enzymic Measurement of Total Cholesterol in Serum," *Clinical Chemistry* 24, no. 1 [January 1978]: 108–14). Through telephone inquiry we determined that the test packs used in the research were filled to the researchers' specifications at the Du Pont plant. This would be in line with the hypothesis that users would find it hard to do this task themselves.

[b]Du Pont commercial chemistries are always sold to the user prepackaged. Technicon commercialized chemistries may be either premixed reagents sold to the user or Technicon-specified formulas the user mixes up in his laboratory as needed.

Commercial Value of User-Developed Innovations

Of course, the practical value manufacturers can expect from shifting user innovation toward, or away from, their products depends on the potential commercial value of such user innovation. Do users really develop product modifications of general commercial interest? We explored this question in the case of both user-developed test methods and user-developed hardware modifications to Technicon and Du Pont clinical chemistry autoanalyzers.

Commercial Value of User-Developed Test Methods

The test methods of most commercial interest to manufacturers of clinical chemistry analyzers are generally those most frequently used (purchased) by clinical chemistry laboratories. To get an indication of the commercial potential of user activity in test development, we decided to explore whether users had played a role in adapting some of these frequently purchased test methods for use on the Technicon and Du Pont brands of autoanalyzer.

Our sample of commercially successful test methods consisted of the 20 most frequently performed clinical chemistry tests.[6] Automated methods for performing 20 of these tests were offered by Technicon and 18 by Du Pont;

TABLE 7–2. Source of Automated Test Methods Offered Commercially by Equipment Manufacturers[a]

	% User	*User*[b]	*Equipment Manufacturer*	*Reagent Manufacturer*	*NA*	*Total*
Du Pont aca	0%	0	18	0	0	18
Technicon SMAC[c]	74	14	4	1	1	20
				Null hypothesis that test method source identical for both brands of analyzers rejected: $p < .0001$ (Fisher exact)		

[a]As explained in the text, the sources of adaptation to automation of the 20 clinical chemistry tests performed with greatest frequency in 1977 were examined. These 20 tests are: albumin; alk phos; calcium; chloride; cholesterol; CPK; creatinine; direct (conjugated) bilirubin; total bilirubin; glucose; SGOT; SGPT; inorganic phos; LDH; potassium; sodium; total protein; triglycerides; urea nitrogen; uric acid. All 20 are offered by Technicon; Du Pont offers all but potassium and sodium.

In the numerous instances in which manufacturers offered different methods for the performance of a given test on their equipment with the passage of time, the method offered for use on the most recently introduced equipment model at the time of the study was the one we selected for inclusion in the sample. At the time of the study, Technicon's latest model was trade named the SMAC High-Speed Computer-Controlled Biochemical Analyzer. Du Pont's latest model at the time of the study was trade named aca (automated clinical analyzer).

[b]The measure used: Did one or more users publish a report of adaptation and clinical use of a given test method on Du Pont or Technicon equipment with publication date prior to the date of commercial introduction of that method (as reported by equipment manufacturer personnel)? Those who performed the adaptation to automation of a test method were coded on the basis of professional affiliation. In the event, all except three innovating users were found to be professionals working in clinical laboratories of nonprofit hospitals. The three exceptions worked in an automated methods laboratory in a Veteran's Administration hospital and were affiliated with that hospital's clinical laboratory.

[c]Some of the user-developed methods offered by Technicon to SMAC users had been developed by users on earlier models of Technicon equipment.

this yielded a sample of 38 adaptations to automation for study. (Because Du Pont and Technicon analyzers have different performance characteristics, the task of developing an analyzer-compatible version of test method x for a Du Pont analyzer is independent of the task of developing an analyzer-compatible version of the same method for a Technicon analyzer.) The innovation history of each sample member was determined through literature searches and structured interviews with manufacturer and user personnel.[7]

In Table 7–2 we see that user-developed adaptations of test methods to automation clearly can be commercially important. For example, 74% of the test methods most frequently used by Technicon customers were first adapted for use on Technicon autoanalyzers by users. In sharp contrast [$p < .0001$], no user-developed adaptations to automation were commercialized by Du Pont. As we saw earlier, the Du Pont equipment was not hospitable to user innovation and attracted very little of it. Users *may* develop product modifications that the manufacturer will find commercially valuable but only if they have an incentive to modify that manufacturer's products.

TABLE 7–3. Du Pont and Technicon Autoanalysis Equipment Innovations

Du Pont autoanalyzer equipment innovations
Basic innovation: Original Du Pont aca autoanalyzer
MAJOR IMPROVEMENTS
Improved computer control
Technicon continuous flow autoanalyzer equipment innovations
Basic innovation: First commercialized model
MAJOR DETECTOR IMPROVEMENTS
Fluorometer
Flame ionization photometer
Ion selective electrodes
MAJOR FLOW CELL IMPROVEMENTS
Smaller volume/adjacent debubbler
Bubble-gated flow cell
MAJOR DIALYSIS IMPROVEMENTS
Shorter flow path/type C membrane
Type H membrane
REDUCTION IN SAMPLE CARRYOVER
Reduced tubing diameter
Air/sample/reagent pump synchronization
Multiple bubble introduction by sample probe
Computer compensation for carryover
OTHER MAJOR IMPROVEMENTS
Multiple channel equipment
Physician-readable chart output

Commercial Value of User-Developed Hardware

We have seen that user-developed test protocols have value for autoanalyzers. What about user-designed modifications to the analyzer hardware? We examined this matter in the case of Technicon and Du Pont analyzers.

Our sample for this test consisted of the first autoanalyzer hardware commercialized by Technicon and Du Pont and all major improvements to that hardware commercialized by each manufacturer over the succeeding years (Table 7–3). We identified these innovations by first talking to manufacturer personnel to identify all hardware changes they had commercialized. Then, expert users and manufacturer personnel helped us determine which of these had resulted in a significant increment in functional utility to the user of the analyzers.

As Table 7–3 shows, we were able to identify 13 Technicon equipment improvement innovations that met our selection criterion, but only 1 such innovation in the instance of Du Pont. (Du Pont equipment, we found, had remained almost unchanged since its commercial introduction.[8]) Data collection to determine the functional sources of innovation for this sample was done by telephone interviewing of expert user and manufacturer personnel.

Since Technicon was the only firm we examined that did commercialize a

TABLE 7–4. Sources of Clinical Chemistry Autoanalyzer Innovations

		Innovation Developed by			
Analyzer Type	*% User*	*User*	*Manufacturer*	*NA*	*Total*
Du Pont aca					
Basic innovation	0%	0	1	0	1
Major improvements	0	0	1	0	1
Technicon continuous flow					
Basic innovation	100	1	0	0	1
Major improvements	46	6	7	0	13
TOTAL		7	9	0	16

significant number of improvements to analyzer hardware, we can only really test whether user-developed hardware innovations sometimes had commercial value in the instance of the Technicon equipment. As Table 7–4 shows, user hardware innovations did in fact have significant commercial value for Technicon. The first clinical chemistry autoanalyzer they produced (in fact it was the first instrument of this type introduced by any firm) was developed by a user. Also, almost half of the hardware improvements Technicon commercialized during the succeeding years were developed by users. In contrast, the basic Du Pont aca and the single major hardware improvement to that analyzer commercialized over the years was developed by Du Pont itself.

Note that the absence of commercialized user-developed hardware modifications for the Du Pont aca is not a consequence of Du Pont policy. Instead, it is likely that users simply did not develop hardware modifications for the aca because this was more costly than modifying functionally similar Technicon equipment. As we mentioned earlier, the Technicon equipment consisted of modules interconnected by plastic tubing. In contrast, the Du Pont analyzer is of a more monolithic design.

Summary

Our data on innovation in clinical chemistry autoanalyzers suggest that the functional source of innovation related to particular products *can* be modified or shifted by actions taken by individual firms.

In the particular sample we have studied, product design appears to be the principal cause of the interbrand difference in the user innovation activity we observed. But any variable that will create shifts in the locus of innovation-related rents, however achieved, should be usable to achieve similar effects. Thus, product manufacturers who wish to increase user innovation affecting their products might offer free equipment or design help to the innovating users they want to encourage. Or, if they want to decrease user innovation, they could refuse to service products users have modified, seal the product physically to hamper user access, or refuse user requests for technical help, circuit diagrams, software source codes, and so on.

Finally, although our study has dealt with a product manufacturer's ability to affect user innovation, the reasoning is perfectly general: Users, suppliers, and even government (for example, through tax policy and/or government purchases and/or government-sponsored research) should also be able to engage in shifting the functional source of innovation if they wish to do so.

Notes

1. Eric von Hippel and Stan N. Finkelstein, "Analysis of Innovation in Automated Clinical Chemistry Analyzers," *Science & Public Policy* 6, no. 1 (February 1979): 24–37.

2. Approximately 44% of the 677 million clinical chemistry tests performed in hospital laboratories in the United States in 1977 were performed on automated clinical chemistry analyzers. In 1975 there were some 14,000 clinical chemistry laboratories in the United States. Some 50% of these were affiliated with hospitals, 30% were affiliated with doctors' offices, and 20% were independent commercial entities. Their aggregate revenues were on the order of $6.2 billion in 1975 and growing at 10% annually (L. H. Smithson, *Overview of the Clinical Laboratory Market* [Menlo Park, Calif.: Stanford Research Institute, n.d.]).

3. IMS America, *Semi-Annual Audit of Laboratory Tests, Hospital Labs,* January–June 1977, July–December 1977 (Ambler, Penn.: IMS America, n.d.). IMS America generates its data by surveying and auditing laboratory records of 204 of the approximately 5800 nonfederal, short-term hospitals in the United States. The sample of hospital laboratories used is stratified by bed size, region, and hospital ownership. (IMS restricts circulation of its data; it is used here by permission of the company.)

4. Technicon offers several models of automated clinical chemistry analyzer, Abbott Laboratories offers two models, and Du Pont one. All models of a given manufacturer are fitted with the same type of reagent proportioning system; however, as a consequence, we will be able to examine the hypothesis by collecting data on analyzer *brands* rather than on specific models of analyzer.

5. To accomplish this MEDLINE was instructed to search for articles that were coded under the subject heading "autoanalysis" and that *also* contained the words "Du Pont," "Technicon," or "Abbott Laboratories" in the article's title and/or abstract. (Although this procedure flags only the subset of research publications that name the autoanalyzer equipment manufacturer in the title and/or abstract, it is reasonable that the ratio of the two types of research usage we are considering will be equal in this subset and in the total population.)

Autoanalysis is the subject heading assigned in the MEDLINE thesaurus (National Library of Medicine, *Medical Subject Headings—Annotated Alphabetical List, 1978* [Springfield, Mass.: U.S. Department of Commerce, National Technical Information Service (No. PB-270-894), 1978]) to research using clinical chemistry autoanalyzers. This thesaurus of standard subject headings is maintained for use by indexers and those wishing to retrieve citations. The MEDLINE system provides access to articles published in most biomedical journals (approximately 3000) from 1964 to 1975 to the present by title, author, subject heading, and textword. Subject headings are assigned to articles by indexers working for the National Library of Medicine as a function of the subject matter dealt with in the article. Textwords are simply any word or combination of words. Users of the system may specify textwords and the system will flag articles containing them in the article title and/or abstract.

6. IMS America, *Semi-Annual Audit of Laboratory Tests, Hospital Labs,* Table 6–2.

7. Work on each case began with a search of the literature for papers related to the test method being examined. Authors whose papers were found germane were contacted and were told that we were interested in exploring the early history of the application of the innovation discussed in their papers to autoanalyzers. We then asked them for the names of fellow experts with user and/or manufacturer and/or other relationships to the innovation who might have a good knowledge of these matters. Finally, we asked these initial contacts for any knowledge they themselves might have on the topic of interest. Individuals identified for us by initial contactees were contacted in turn, and the process repeated until we felt we had the well-documented information we needed.

We found that FDA-required product labeling was an especially useful data source for all sampled innovations. (Product labeling is U.S. Food and Drug Administration [FDA] terminology for methods-related information suppliers of clinical test chemistry methods must make available to their customers. Among other things, product labeling contains references to research behind those methods.)

8. Several equipment changes to the Du Pont aca are listed in B. W. Perry, et al., *A Field Evaluation of the Du Pont Automatic Clinical Analyzer* (Wilmington, Del.: Du Pont, n.d.; 2nd printing, January 1978). We did not include these changes in our sample because they were made prior to the commercial introduction of that analyzer. If we had included them, they would not have changed our finding that users do not develop equipment improvement innovations for the aca. Although the monograph authors were users at the University of Alabama Medical Center in Birmingham, the equipment problems uncovered by their evaluation work were rectified by changes developed by Du Pont personnel.

8

Predicting the Source of Innovation: Lead Users

The experiment I report on in this chapter involves predicting the source of innovation, user innovation in this instance. The specific context of the work addresses an important problem facing industrial and consumer marketing research: How can one accurately determine user needs for new products (processes and services) in fields that are rapidly changing such as those touched by high technology?

I begin by exploring the difficulty faced by marketing research in more depth. Then, I spell out the lead user methodology I have proposed as a solution.[1] Finally, I describe a first application of the method.[2]

Root of the Problem: Marketing Research Constrained by User Experience

Users selected to provide input data to consumer and industrial market analysis have an important limitation: Their insights into new product (and process and service) needs and potential solutions are constrained by their real-world experience. Users steeped in the present are, thus, unlikely to generate novel product concepts that conflict with the familiar.

The notion that familiarity with existing product attributes and uses interferes with an individual's ability to conceive of novel attributes and uses is strongly supported by research into problem solving (Table 8–1). We see that experimental subjects familiar with a complicated problem-solving strategy are unlikely to devise a simpler one when this is appropriate.[3] Also, and germane to our present discussion, we see that subjects who use an object or see it used in a familiar way are strongly blocked from using that object in a novel way.[4] Furthermore, the more recently objects or problem-solving strategies have been used in a familiar way, the more difficult subjects find it to employ them in a novel way.[5] Finally, we see that the same effect is displayed in the real world, where the success of a research group in solving a new problem is shown to depend on whether solutions it has used in the past will fit that new problem.[6] These studies thus suggest that typical users of existing products—the type of user-evaluators customarily chosen in market research—are poorly situated with regard to the difficult problem-solving tasks associated with assessing unfamiliar product and process needs.

As illustration consider the difficult problem-solving steps potential users must go through when asked to evaluate their need for a proposed new product. Since individual industrial and consumer products are only components in larger usage patterns that may involve many products and since a change in one component can change perceptions of, and needs for, some or all other products in that pattern, users must first identify their existing multiproduct usage patterns in which the new product might play a role. Then, they must evaluate the new product's potential contribution to these. (E.g., a change in the operating characteristics of a computer may allow users to solve new problem types if they also make changes in software and perhaps in other, related products and practices.) Next, users must invent or select the new (to them) usage patterns that the proposed new product makes possible for the first time and then evaluate the utility of the product in these patterns. Finally, since substitutes exist for many multiproduct usage patterns (e.g., many forms of problem analysis are available in addition to the novel ones made possible by a new computer), the user must estimate how the new possibilities presented by the proposed new product will compete (or fail to compete) with existing options. This problem-solving task is clearly a very difficult one, particularly for typical users of existing products whose familiarity with existing products and uses interferes with their ability to conceive of novel products and uses when invited to do so.

The constraint of users to the familiar pertains even in the instance of sophisticated marketing research techniques such as multiattribute mapping of product perceptions and preferences.[7] Multiattribute (multidimensional) marketing research methods, for example, describe users' (buyers') perception of new and existing products in terms of a number of attributes (dimensions). If and as a complete list of attributes is available for a given product category, the users' perception of any particular product in the category can be expressed in terms of the amount of each attribute they perceive it to contain, and the difference between any two products in the category can be expressed as the difference in their attribute profiles. Similarly, users' preferences for existing and proposed products in a category can in principle be built up from their perceptions of the importance and desirability of each of the component product attributes.

Although these methods frame user perceptions and preferences in terms of attributes, they do not offer a means of going beyond the experience of those interviewed. First, for reasons discussed earlier, users are not well positioned to accurately evaluate novel product attributes or accurately quantify familiar product attributes that lie outside the range of their real-world experience. Second, and more specific to these techniques, there is no mechanism to induce users to identify all product attributes potentially relevant to a product

TABLE 8–1. The Effect of Prior Experience on Users' Ability to Generate or Evaluate Novel Product Possibilities

Study	*Nature of Research*	*Impact of Prior Experience on Ability to Solve Problems*
1. Luchins (1942)	Two groups of subjects ($n =$) were given a series of problems involving water jars, for example, "If you have jars of capacity *A*, *B*, and *C*, how can you pour water from one to the other so as to arrive at amount *D*?" Subject group 1 was given five problems solvable by formula, $B - A - 2C = D$. Next, both groups were given problems solvable by that formula *or* by a simpler one (e.g., $B - C = D$).	81% of experimental subjects who had previously learned a complex solution to a problem type applied it to cases where a simple solution would do. No control group subjects did so ($p =$ NA).[a]
2. Duncker (1945)	The ability to use familiar objects in an unfamiliar way was tested by creating five problems that could only be solved by that means. (E.g., one problem could be solved only if subjects bent a paper clip provided them and used it as a hook.) Subjects were divided into two groups. One group of problem solvers saw the crucial object being used in a familiar way (e.g., the paper clip holding papers), the other did not (e.g., the paper clip was simply lying on a table unused).	Subjects were much more likely to solve problems requiring the use of familiar objects in unfamiliar ways if they had not been shown the familiar use just prior to their problem-solving attempt. Duncker called this effect "functional fixedness" ($n = 14$; $p =$ NA).[a]
3. Birch and Rabinowitz (1951)	Replication of Duncker, above.	Duncker's findings confirmed ($n = 25$; $p < .05$).
4. Adamson (1952)	Replication of Duncker, above.	Duncker's findings confirmed ($n = 57$; $p < .01$).
5. Adamson and Taylor (1954)	The variation of functional fixedness with time was observed by the following procedure. First, subjects were allowed to use a familiar object in a familiar way. Next, varying amounts of time were allowed to elapse before subjects were invited to solve a problem by using the object in an *un*familiar way.	If a subject uses an object in a familiar way, he or she is partially blocked from using it in a novel way ($n = 32$; $p < .02$). This blocking effect decreases over time.
6. Allen and Marquis (1964)	Government agencies often buy R & D services through a request for proposal (RFP) that states the problem to be solved. Interested	Bidders were significantly more likely to propose a successful task approach if they had prior experience with that approach only

	bidders respond with proposals that outline their planned solutions to the problem and its component tasks. In this research the relative success of eight bidders' approaches to the component tasks contained in two RFPs was judged by the agency buying the research ($n = 26$). Success was then compared to prior research experience of bidding laboratories.	rather than prior experience with inappropriate approaches only.

Source: Eric von Hippel, "Lead Users: A Source of Novel Product Concepts," *Management Science* 32, no. 7 (July 1986), 794—95. Copyright 1986, The Institute of Management Sciences, 290 Westminster Street, Providence, Rhode Island 02903.

[a]This relatively early study showed a strong effect but did not provide a significance calculation or present data in a form that would allow one to be determined without ambiguity.

category, especially attributes that are currently not present in any extant category member. To illustrate this point, consider two types of such methods, similarity-dissimilarity ranking and focus groups.

In similarity-dissimilarity ranking, data regarding the perceptual dimensions by which users characterize a product category are generated by asking a sample of users to compare products in that category and assess them in terms of their similarity and dissimilarity. In some variants of the method, the user interviewee specifies the ways in which the products are similar or different. In others, the user simply provides similarity and difference rankings, and the market researcher determines—through his personal knowledge of the product type in question, its function, the marketplace, and so on—the important perceptual dimensions that must be motivating the user rankings obtained.

The similarity-dissimilarity method clearly depends heavily on an analyst's qualitative ability to interpret the data and correctly identify all the critical dimensions. Moreover, by its nature, this method can only explore perceptions derived from attributes that exist in, or are associated with, the products being compared. Thus, if a group of evaluators is invited to compare a set of cameras and none has a particular feature—say, instant developing—then the possible utility of this feature would not be incorporated in the perceptual dimensions generated. That is, the method would have been blind to the possible value of instant developing prior to Edwin Land's invention of the Polaroid camera.

In focus group methods, market researchers assemble a group of users familiar with a product category for a qualitative discussion of perhaps two hours' duration. The topic for the focus group, which is set by the market researcher, may be relatively narrow (e.g., users' perceptions of *x* brand) or somewhat broader (e.g., camera users' perceptions of the photographic experience). The ensuing discussion is recorded, transcribed, and later reviewed by the researcher, whose task it is to identify the important product attributes that have implicitly or explicitly surfaced during the conversation. Clearly, as with similarity-dissimilarity ranking, the utility of information derived from focus group methods depends heavily on the individual analyst's ability to accurately and completely abstract from the interview data the attributes users feel important in products.

In principle, however, the focus group technique need not be limited to only identifying attributes already present in existing products, even if the discussion is nominally focused on these. For example, a topic that extends the boundaries of discussion beyond a given product to a larger framework could identify attributes not present in any extant product in a category under study. If discussion of the broad topic mentioned earlier, camera users' perceptions of the photographic experience, brought out dissatisfaction with the time lag between picture taking and receipt of the finished photograph, the analyst would be in possession of information that could induce him to identify an attribute not present in any camera prior to Land's invention, instant film development, as a novel and potentially important attribute.

But how likely is it that an analyst will take this creative step? And, more generally, how likely is it that either method discussed above, similarity-dissimilarity ranking or focus groups, will be used to identify attributes not present in extant products of the type being studied, much less a complete list of all relevant attributes? Neither method contains an effective mechanism to encourage this outcome, and discussions with practitioners indicate that in present-day practice, identification of any novel attribute is unlikely.

Finally, both of these methods conventionally focus on familiar product categories. This restriction, necessary to limit the number of attributes that completely describe a product type to a manageable number, also tends to limit market research interviewees to attributes that fit products within the frame of existing product categories. Modes of transportation, for example, logically shade off into communication products as partial substitutes ("I can drive over to talk to him, or I can phone"), into housing and entertainment products ("We can buy a summer house, or go camping in my recreational vehicle"), indeed, into many other of life's activities. But since a complete description of life cannot be compressed into 25 attribute scales, the analysis is constrained to a narrower—usually conventional and familiar—product category or topic. This has the effect of rendering any promising and novel cross-category new product attributes less visible to the methods I have discussed.

In sum, then, we see that marketing researchers face serious difficulties if they attempt to determine new product needs falling outside of the real-world experience of the users they analyze.

Lead Users as a Solution

In many product categories, the constraint of users to the familiar does not lessen the ability of marketing research to evaluate needs for new products by analyzing typical users. In the relatively slow-moving world of steels and autos, for example, new models often do not differ radically from their immediate predecessors. Therefore, even the "new" is reasonably familiar and the typical user can thus play a valuable role in the development of new products.

In contrast, in high technology industries, the world moves so rapidly that the related real-world experience of ordinary users is often rendered obsolete by the time a product is developed or during the time of its projected commercial lifetime. For such industries I propose that lead users who *do* have real-life experience with novel product or process needs are essential to accurate marketing research. Although the insights of lead users are as constrained to the familiar as those of other users, lead users are familiar with conditions that lie in the future for most—and, so, are in a position to provide accurate data on needs related to such future conditions.

I define lead users of a novel or enhanced product, process, or service as those who display two characteristics with respect to it:

1. Lead users face needs that will be general in a marketplace, but they face them months or years before the bulk of that marketplace encounters them, *and*
2. Lead users are positioned to benefit significantly by obtaining a solution to those needs.

Thus, a manufacturing firm with a current strong need for a process innovation that many manufacturers will need in two years' time would fit the definition of lead user with respect to that process.

Each of the two lead user characteristics provides an independent contribution to the type of new product need and solution data such users are hypothesized to possess. The first specifies that a lead user will possess the particular real-world experience manufacturers must analyze if they are to accurately understand the needs the bulk of the market will have tomorrow. Users "at the front of the trend" typically exist simply because important new technologies, products, tastes, and other factors related to new product opportunities typically diffuse through a society, often over many years, rather than impact all members simultaneously.[8]

The second lead user characteristic is a direct application of the hypothesis we have focused on in this book, and assumes it correct: Users who expect high rents from a solution to a need under study should (I reason) have been driven by these expectations to attempt to solve their need. This work in turn will have produced insight into the need and perhaps useful solutions that will be of value to inquiring market researchers.

In sum, then, lead users are users whose present strong needs will become general in a marketplace months or years in the future. Since lead users are familiar with conditions that lie in the future for most others, I hypothesize that they can serve as a need-forecasting laboratory for marketing research. Moreover, since lead users often attempt to fill the need they experience, I hypothesize that they can provide valuable new product concept and design data to inquiring manufacturers in addition to need data.

Testing the Method

Glen Urban and I, with the able assistance of our student, David Israel-Rosen, have conducted a prototype lead user market research study in the rapidly changing field of computer-aided-design (CAD) products. (Over 40 firms compete in the $1 billion market for CAD hardware and software. This market grew at over 35% per year over the period 1982 to 1986 and the forecast is for continued growth at this rate for the next several years.) Within the CAD field, we decided to specifically focus on CAD systems used to design the printed circuit (PC) boards used in electronic products, PC–CAD.

Printed circuit boards hold integrated circuit chips and other electronic components and interconnect these into functioning circuits. PC–CAD systems help engineers convert circuit specifications into detailed printed circuit board designs. The design steps that are, or can be, aided by PC–CAD include component placement, signal routing (interconnections), editing and checking, documentation, and interfacing to manufacturing. The software required to perform these tasks is quite complex and includes placement and routing algorithms and sophisticated graphics. Some PC–CAD manufacturers sell only such software, whereas others sell systems that include both specialized computers and software. (Important suppliers of PC–CAD in 1985 included IBM, Computervision, Redac, Calma, Scicards, and Telesis.)

The method Urban and I used to identify lead users and test the value of the data they possess in the PC–CAD field involved four major steps: (1) identify an important market or technical trend, (2) identify lead users with respect to that trend, (3) analyze lead user data, and (4) test lead user data on ordinary users. I will discuss each in turn.

Identifying an Important Trend

Lead users are defined as being in advance of the market with respect to a given important dimension that is changing over time. Therefore, before one can identify lead users in a given product category of interest, one must specify the underlying trend on which these users have a leading position.

To identify an "important" trend in PC–CAD, we sought out a number of expert users. We identified these by telephoning managers of the PC–CAD groups of a number of firms in the Boston area and asking each: "Whom do you regard as the engineer most expert in PC–CAD in your firm?" "Whom in your company do group members turn to when they face difficult PC–CAD problems?"[9] After our discussions with expert users, it was qualitatively clear to us that an increase in the density with which chips and circuits are placed on a board was, and would continue to be, a very important trend in the PC–CAD field. Historical data showed that board density had in fact been steadily increasing over a number of years. And the value of continuing increases in density was clear. An increase in density means that it is possible to mount more electronic components on a given size printed circuit board. This in turn translates directly into an ability to lower costs (less material is used), to decreased product size, and to increased speed of circuit operation (signals between components travel shorter distances when board density is higher).

Very possibly, other equally important trends exist in the field that would reward analysis, but we decided to focus on this single trend in our study.

Identifying Lead Users

To identify lead users of PC–CAD systems capable of designing high-density printed circuit boards, we had to identify that subset of users: (1) who were designing very high-density boards now and (2) who were positioned to gain especially high benefit from increases in board density. We decided to use a formal telephone-screening questionnaire to accomplish this task, and we strove to design one that contained objective indicators of these two hypothesized lead user characteristics.

Printed circuit board density can be increased in a number of ways and each offers an objective means of determining a respondent's position on the trend toward higher density. First, the number of layers of printed wiring in a printed circuit board can be increased. (Early boards contained only 1 or 2 layers but now some manufacturers are designing boards with 20 or more layers.) Second, the size of electronic components can be decreased. (A recent important technique for achieving this is surface-mounted devices that are soldered directly to the surface of a printed circuit board.) Finally, the printed wires, vias, that interconnect the electronic components on a board can be made narrower and packed more closely. Questions regarding each of these density-related attributes were included in our questionnaire.

Next, we assessed the level of benefit a respondent might expect to gain by improvements in PC–CAD by means of several questions. First, we asked about users' level of satisfaction with existing PC–CAD equipment, assuming that high dissatisfaction would indicate expected high benefit from improvements. Second, we asked whether respondents had developed and built their own PC–CAD systems rather than buy the commercially available systems such as those offered by IBM or Computervision. (We assumed, as we noted previously, that users who make such innovation investments do so because they expect high benefit from resulting PC–CAD system improvements.) Finally, we asked respondents whether they thought their firms were innovators in the field of PC–CAD.

The PC–CAD users interviewed were restricted to U.S. firms and selected from two sources: A list of members of the relevant professional engineering association (IPCA) and a list of current and potential customers provided by a cooperating supplier. Interviewees were selected from both lists at random. We contacted approximately 178 qualified respondents and had them answer the questions on the phone or by mail if they preferred. The cooperation rate was good: 136 screening questionnaires were completed. One third of these

TABLE 8–2. Cluster Analyses Show User Group with Hypothesized Lead User Characteristics

	Two-Cluster Solution		*Three-Cluster Solution*		
	Lead User	*Nonlead User*	*Lead User*	*Non-lead (A)*	*Non-lead (B)*
Indicators of user position on PC–CAD density trend					
Use surface mount?	87%	56%	85%	7%	100%
Average line width (mils)	11	15	11	17	13
Average layers (number)	7.1	4.0	6.8	4.2	4.4
Indicators of user-expected benefit from PC–CAD improvement					
Satisfaction[a]	4.1	5.3	4.1	5.2	5.2
Indicators of related user innovation					
Build own PC–CAD?	87 %	1 %	100 %	0 %	0 %
Innovativeness[b]	3.3	2.4	3.2	2.1	2.8
First use of CAD (year)	1973	1980	1973	1980	1979
Number in cluster	38	98	33	46	57

[a]7-point scale—high value more satisfied.
[b]4-point scale—high value more innovative.

were completed by engineers or designers, one third by CAD or printed circuit board managers, 26% by general engineering managers, and 8% by corporate officers.

Simple inspection of the screening questionnaire responses showed that fully 23% of all responding user firms had developed their own in-house PC–CAD hardware and software systems. This high proportion of user-innovators that we found in our sample is probably characteristic of the general population of PC–CAD users. Our sample was well dispersed across the self-stated scale with respect to innovativeness: 24% indicated they were on the leading edge of technology, 38% up-to-date, 25% in the mainstream, and 13% adopting only after the technology is clearly established. This self-perception is supported by objective behavior with respect to the alacrity with which our respondents adopted PC–CAD.

We next conducted a cluster analysis of screening questionnaire data relating to the hypothesized lead user characteristics in an attempt to identify a lead user group. The two- and three-cluster solutions are shown in Table 8–2.

Note that these analyses do, indeed, clearly indicate a group of respondents who combine the two hypothesized attributes of lead users and that, effectively, all of the PC–CAD product innovation is reported by the lead user group.

In the two-cluster solution, what we term the lead user cluster is, first, ahead of nonlead users in the trend toward higher density. That is, lead users report more use of surface-mounted components, use of narrower lines, and use of more layers than do members of the nonlead cluster. Second, lead users appear to expect higher benefit from PC–CAD innovations that would allow them even further progress. That is, they report less satisfaction with their existing PC–CAD systems (4.1 vs. 5.3, with higher values indicating satisfaction). Strikingly, 87% of respondents in the lead user group report building their own PC–CAD system (vs. only 1% of nonlead users) in order to obtain improved PC–CAD system performance.[10] Lead users also judged themselves to be more innovative (3.3 vs. 2.4 on the four-statement scale with higher values more innovative), and they were in fact earlier adopters of PC–CAD than were nonlead users.

Note that 28% of our respondents are classified in this lead user cluster. The two clusters explained 24% of the variation in the data.

In the three-cluster solution the lead user group was nearly unchanged, but the nonlead group was separated into two subgroups. Nonlead group *A* had the lowest use of surface-mounted components, the widest line widths, the fewest layers, and the latest year of adoption, and it rated itself as lowest on adoption of innovations. Nonlead group *B* also differed from the lead user group in the expected ways, except for one anomalous result: Nonlead group *B* showed a higher usage of surface-mounted components than did the lead user group. In the three-cluster solution 37% of the variation is explained by cluster membership.

A discriminant analysis indicated that building one's own system was the most important indicator of membership in the lead user cluster. (The discriminant analysis had 95.6% correct classification of cluster membership. The standardized discriminant function of coefficients were: build own .94, self-stated innovativeness .27, average layers .25, satisfaction −.23, year of adoption −.16, surface mounting .152.)

Analyzing Lead User Insights

The next step in our analysis was to select a small sample of the lead users identified in our cluster analysis to participate in a group discussion to develop one or more concepts for improved PC–CAD systems. Experts from five lead user firms that had facilities located near MIT were recruited for this group. The firms represented were Raytheon, DEC, Bell Laboratories, Honeywell, and Teradyne. Four of these five firms had built their own PC–CAD systems. All were working in high-density (many layers and narrow lines) applications and had adopted the CAD technology early.

The task set for this group was to specify the best PC–CAD system for laying out high-density digital boards that could be built with current technology. (To guard against the inclusion of "dream" features impossible to implement, we conservatively allowed the concept the group developed to include only features that one or more of them had already implemented in their own organizations. No one firm had implemented all aspects of the concept, however.)

The PC–CAD system concept developed by our lead user creative group integrated the output of PC–CAD with numerically controlled printed circuit board manufacturing machines; had easy input interfaces (e.g., block diagrams, interactive graphics, icon menus); and stored data centrally with access by all systems. It also provided full functional and environmental simulation (e.g., electrical, mechanical, and thermal) of the board being designed and could design boards of up to 20 layers, route thin lines, and properly locate surface-mounted devices on the board.

Testing Product Concept Perceptions and Preferences

From the point of view of marketing research, new product need data and new product solutions from lead users are only interesting if they are preferred by the general marketplace.

To test this matter, we decided to determine PC–CAD user preferences for four system concepts: the system concept developed by the lead user group, each user's own in-house PC–CAD system, the best commercial PC–CAD system available at the time of the study (as determined by a PC–CAD system manufacturer's competitive analysis), and a system for laying out curved printed circuit boards. (This last was a description of a special-purpose system that one lead user had designed in-house to lay out boards curved into three-dimensional shapes. This is a useful attribute if one is trying to fit boards into the oddly shaped spaces inside some very compact products, but most users would have no practical use for it. In our analysis of preference, we think user response to this concept can serve to flag any respondent tendency to prefer systems based on system exotica rather than practical value in use.)

To obtain user preference data regarding our four PC–CAD system concepts, we designed a new questionnaire that contained measures of both perception and preference. First, respondents were asked to rate their current PC–CAD system on 17 attribute scales. (These were generated by a separate sample of users through triad comparisons of alternate systems, open-ended interviews, and technical analysis. Each scale was presented to respondents in the form of a five-point agree-disagree judgment based on a statement such as "my system is easy to customize."[11] Next, each respondent was invited to read a one-page description of each of the three concepts we had generated (labeled simply, *J, K,* and *L*) and rate them on the same scales. All concepts were described as having an identical price of $150,000 for a complete hardware and software workstation system able to support four users. Next, rank-order preference and constant-sum paired comparison judgments were requested for the three concepts and the existing system. Finally, probability-of-purchase measures on an 11-point Juster scale were collected for each concept at the base price of $150,000, with alternate prices of $100,000 and $200,000.

Our second questionnaire was sent to 173 users (the 178 respondents who

TABLE 8–3 Test of All Respondents' Preferences Among Four Alternative PC–CAD System Concepts

PC–CAD Concept	*% First Choice*	*Constant Sum*[a]	*Average Probability of Purchase*
Lead user group concept	78.6	2.60	51.7
Respondents' current PC–CAD	9.8	1.87	[b]
Best system commercially available	4.9	0.95	20.0
User system for special application	6.5	0.77	26.0

[a]Warren S. Torgerson, *Theory and Methods of Scaling* (New York: Wiley, 1958).
[b]Probability of purchase only collected across concepts.

qualified in the screening survey less the 5 user firms in the creative group). Respondents were called by phone to inform them that a questionnaire had been sent. After telephone follow-up and a second mailing of the questionnaire, 71 complete or near-complete responses were obtained (41%) and the following analyses are based on these.[12]

Lead User Concept Preferred

As can be seen from Table 8–3, our analysis of the concept questionnaire showed that respondents strongly preferred the lead user group PC–CAD system concept over the three others presented to them: 78.6% of the sample selected the lead user creative group concept as their first choice. The constant sum scaled-preference value was 2.60 for the concept developed by the lead user group. This was 35% greater than users' preference for their own current system and more than twice as great as the preference for the most advanced existing commercially available product offering.

The concept created by the lead user group was also generally preferred by users over their existing systems (significant at the 10% level based on the preference measures: $t = 12$ for proportion first choice and $t = 2.1$ for constant sum). And, the lead user group concept was significantly preferred over the special application user system developed to lay out curved boards. (The lead user concept was significantly better than the user-developed special application system on all measures at the 10% level ($t = 12.3$ for first choice, $t = 7.9$ for preference, and $t = 8.5$ for probability).[13]

Respondents maintained their preferences for the lead user concept even when it was priced higher than competing concepts. The effects of price were investigated through the probability of purchase measures collected at three prices for each concept. For the lead user concept, probability of purchase

TABLE 8–4. Concept Preferences of Lead Versus Nonlead Users

PC–CAD Concept	% First Choice		Constant Sum		Average Probability of Purchase	
	Lead	Nonlead	Lead	Nonlead	Lead	Nonlead
Lead user group concept	92.3%	80.5%	3.20	2.37	53.1	51.2
Respondents' current PC–CAD	7.7	11.1	2.64	1.56	0	0
Best system commercially available	0	2.8	0.67	1.06	10.2	23.9
User system for special application	0	5.6	0.52	0.87	16.3	29.9

increases from 52.3% to 63.0% when the price is decreased from $150,000 to $100,000 ($t = 2.3$) and drops to 37.7% when the price is increased to $200,000. Probability of purchase of the lead user concept was significantly higher at all price levels (t greater than 4.4 in all paired comparisons), and it was preferred to the best available concept even when the specified price was twice as high as that of competing concepts. All three concepts displayed the same proportionate change in purchase probability as the price was changed from its base level of $150,000. The probability measures indicate substantial price sensitivity and provide a convergent measure on the attractiveness of the concept based on lead user solution content.

Similarity of Lead and Nonlead User Preferences

The needs of today's lead users are typically not precisely the same as the needs of the users who will make up a major share of tomorrow's predicted market. Indeed, the literature on diffusion suggests that in general the early adopters of a novel product or practice differ in significant ways from the bulk of the users who follow them.[14] However, in this instance, as Table 8–4 shows, the product concept preferences of lead users and nonlead users were very similar.

A comparison of the way in which lead and nonlead users evaluated PC–CAD systems showed that this similarity of preference was deep-seated. An examination of the PC–CAD attribute ratings and factor analyses derived from each group showed five factors in each that explained the same amount of variation (67.8 for lead users and 67.7 for nonlead users). The factor loadings were also similar for the two groups, and their interpretation suggested the same dimension labels. Also, analysis showed that each group placed a similar degree of importance on each dimension.[15]

Discussion

From the point of view of marketing research, I think that the results of this first test of a lead user method must be seen as encouraging. Lead users with the hypothesized characteristics were clearly identified; a novel product concept was created based on lead user insights and problem-solving activities; and the lead user concept was judged to be superior to currently available alternatives by a separate sample of lead and nonlead users. I should point out, however, that the high level of actual product innovation found among lead users of PC–CAD can only be expected in product areas where the rents expected by such users are sufficient to induce user innovation. Where expected user benefit is less, need data available from lead users should still be more accurate and richer in "solution content" than data from nonlead users, but it may not include prototype products such as those we have observed in the study of PC–CAD.

From the point of view of my underlying hypothesis regarding the ability to predict the sources of innovation on the basis of innovators' related expectations of rent, I think the lead user application has also shown very encouraging results. Users who identified themselves as dissatisfied with existing products were shown more likely to be involved in developing new ones responsive to their need.

Notes

1. Eric von Hippel, "Lead Users: A Source of Novel Product Concepts," *Management Science* 32, no. 7 (July 1986): 791–805.

2. Glen L. Urban and Eric von Hippel, "Lead User Analyses for the Development of New Industrial Products" (MIT Sloan School of Management Working Paper No. 1797–86) (Cambridge, Mass., June 1986), and *Management Science* (forthcoming).

3. A. S. Luchins, "Mechanization in Problem-Solving: The Effect of *Einstellung*," *Psychological Monographs* 54 (1942).

4. Karl Duncker, "On Problem Solving," trans. Lynne S. Lees, *Psychological Monographs* 58, no. 5 (1945); H. G. Birch and H. J. Rabinowitz, "The Negative Effect of Previous Experience on Productive Thinking," *Journal of Experimental Psychology* 41 (1951): 121–26; and R. E. Adamson, "Functional Fixedness as Related to Problem Solving: A Repetition of Three Experiments," *Journal of Experimental Psychology* 44 (1952): 288–91.

5. R. E. Adamson and D. W. Taylor, "Functional Fixedness as Related to Elapsed Time and to Set," *Journal of Experimental Psychology* 47 (1954): 122–26.

6. T. J. Allen and D. G. Marquis, "Positive and Negative Biasing Sets: The Effects of Prior Experience on Research Performance," *IEEE Transactions on Engineering Management* EM–11, no. 4 (December 1964): 158–61.

7. Alvin J. Silk and Glen L. Urban, "Pre-Test-Market Evaluation of New Packaged Goods: A Model and Measurement Methodology," *Journal of Marketing Research* 15, no. 2 (May 1978): 171–91; Allan D. Shocker and V. Srinivasan, "Multiattri-

bute Approaches for Product Concept Evaluation and Generation: A Critical Review," *Journal of Marketing Research* 16, no. 2 (May 1979): 159–80.

8. For example, when Edwin Mansfield (*The Economics of Technological Change* [New York: Norton, 1968], 134–35) explored the rate of diffusion of 12 very important industrial goods innovations into major firms in the bituminous coal, iron and steel, brewing, and railroad industries, he found that in 75% of the cases it took over 20 years for complete diffusion of these innovations to major firms. Accordingly, some users of these innovations could be found far in advance of the general market.

9. PC–CAD system purchase decisions are made primarily by the final users in the engineering department responsible for CAD design of boards. In this study we interviewed only these dominant influencers to find concepts and test them. If the purchase decision process had been more diffuse, it would have been appropriate to include other important decision participants in our data collection.

10. The innovating users reported that their goal was to achieve better performance than commercially available products could provide in several areas: high routing density, faster turnaround time to meet market demands, better compatibility to manufacturing, interfaces to other graphics and mechanical CAD systems, and improved ease of use for less experienced users.

11. The 17 attributes were: ease of customization, integration with other CAD systems, completeness of features, integration with manufacturing, maintenance, upgrading, learning, ease of use, power, design time, enough layers, high-density boards, manufacturable designs, reliability, placing and routing capabilities, high value, and updating capability.

12. There were 94 individuals (55%) who actually returned the questionnaire, but only 71 returns were judged complete enough to use. This subset consists of 61 respondents who completed all items on both the screening and concept questionnaires, and an additional 10 who completed all items except the constant-sum paired comparison allocations.

13. As part of our analysis we tested for potential nonresponse bias by comparing early and later returns and found none. Returns from the first 41% of respondents showed 77% first choice for the creative group concepts and the last 59% showed 71% first choice. The differences between the early and later returns were not significant at the 10% level (t = .15). Thus, there was no evidence of a nonresponse bias. A possible demand-effect bias toward the lead user group concept could have been present, but the low preferences for the best available product and the curved board concepts argues against it. All concepts were presented in a similar format with labels of concept *J, K,* and *L* (the lead user group concept was labeled *K*). We did not expect any differential bias toward concept *K*.

14. Everett M. Rogers with F. Floyd Shoemaker, *Communication of Innovations: A Cross-Cultural Approach,* 2nd ed. (New York: Free Press, 1971).

15. Urban and von Hippel, "Lead User Analyses for the Development of New Industrial Products."

9

Epilogue: Applications for Innovation Management

In this book, I have presented a view of innovation as a process that is predictably distributed across innovation users, manufacturers, suppliers, and others. To date, I have only explored some of the attributes of this distributed innovation process. Nevertheless, there is no reason for interested innovators to wait for further research: I believe that much of what we know now can be immediately applied to improve innovation management.

In the few pages that follow, I will suggest practical steps for (1) determining one's innovation process role and (2) organizing for it. Finally, I will give an example of system-level strategy that will serve to give the reader a feeling for that concept. I very much hope that interested practitioners will experiment with, and further develop, the possibilities I suggest here. After all, user-developed improvements to innovation practice can offer high rewards to successful user-innovators!

Identifying an Innovation Process Role

When managers are freed from the conventional wisdom of innovation as a preserve of the manufacturer, they can assess their own firm's most efficient role in the innovation process.

On the basis of what we have learned in this book about predicting the sources of innovation, it would seem reasonable for a firm to think that a needed innovation might be available from others if one's own expected rent from innovation is much lower than the rent others may expect. As we saw in chapter 5, it is not necessarily easy to estimate how expectations of innovation benefit are distributed. Happily, however, innovation managers need not make a careful estimate to achieve their purpose. They need only apply the underlying reasoning of chapter 5 to get a rough idea of where innovation is likely to lie, and then they can quickly test their estimates' correctness by contacting likely innovators. If interesting work is in fact going on and if the information is not confidential, they will quickly find the right people to talk to.

In general, feel free to combine the ideas presented in this book with your own industry insight to go beyond the specific cases I have discussed. If you are interested in the sources of innovation for services rather than the products and processes I have explored to date, go ahead and explore—the same principles apply. If you think that firms other than users, manufacturers, or suppliers may innovate in your industry (distributors, for example), simply include these in your assessments just as you would any other likely functional source of innovation. If cooperation in innovation with others (for example, your customers) is of interest, estimate the innovation-related rents that the combination of firms you propose might expect.

When making a search for useful innovation-related activity, one should keep in mind that the most likely source of innovation is dependent on the likely distribution of innovation-related rents in the *precise* category of product, process, or service being considered. Thus, one may estimate that computer firms will be the likely source of many innovations in general-purpose software such as operating systems. At the same time, however, it would be reasonable to expect users to be the most likely source of innovations in specialized applications. Also one should keep in mind that the locus of maximum appropriable rents for an innovation type will not necessarily be one's own firm's users, manufacturers, and/or suppliers. For example, auto firms might sometimes look to another industry entirely (e.g., aerospace) for information on useful materials innovations.

Organizing for an Innovation Process Role

In my experience, it is relatively easy for interested analysts in a firm to determine that firm's appropriate role(s) in the distributed innovation process. It is much harder, however, to take the next step and make organizational changes that the firm may need to play a new role effectively. This is because the design and staffing of any firm's innovation-related activities inevitably contain strong implicit biases about the source of innovation.

Let me illustrate the problem in the context of the manufacturer-as-innovator bias that often exists in the links a product manufacturer establishes with product users. Such links are currently of great concern to manufacturers who correctly see "getting close to the customer" as essential to successful innovation.[1]

First, consider the link, known as field service, between a product manufacturer and its customers. The job of field servicepersons is to go to customer sites and maintain and repair the products a manufacturer has sold. They are equipped with the parts and the diagnostic manuals and equipment needed to maintain and repair standard company products. And, typically, their performance is evaluated on the basis of measures such as time at customer site, which are based on the estimated time to complete standard repairs if the work is done efficiently.

If, in the course of work at a customer site, a field serviceperson working under these conditions comes across a user-modified product, his reaction might well be strongly negative—even if the modification is obviously useful and potentially valuable to the firm and other users. The reason: He probably cannot fix a modified machine rapidly. Possibly he cannot repair it at all with the standard replacement parts in his kit. And in such cases he cannot do his job and may well receive a poor evaluation. Obviously, the conditions described create a systematic manufacturer-as-innovator bias in field service that may or may not be visible to, or intended by, the manager who set up the system.

Another very important link between product manufacturers and their customers is the sales department. Industrial product salespeople, especially, spend much of their time at customer sites and, so, should be in a good position to obtain information on promising user new product needs, ideas, and prototype solutions. But sales departments are typically not staffed with people trained to do this job; the commission and incentive schemes operating on them may well reward only sales of existing products. As a result, salespeople may have no incentive to learn about user developments that might have potential as commercial products. Instead, they may have a positive incentive to deflect any discussion with the customer away from user-developed products and toward the question, "What can I sell you of my present products?"

Next, a product manufacturer's marketing research group is obviously an important link to users' new product needs. But a marketing research group may also contain a manufacturer-as-innovator bias. Traditionally, the task of such a group has been specified as the collection and analysis of data on user needs. Under such ground rules, it often does not occur to anyone to seek data from users on possible sources of new product solution data as well as need data.

Of course, organizational barriers to user solution data do not necessarily end even after the information enters the firm. A firm's R & D group, for example, may well regard such information with a dubious eye. And, given typical incentives and staffing patterns such a reaction, too, is perfectly logical. Note that R & D groups are often staffed with people who are trained to develop new products and processes in-house and are rewarded for this task. Given such a context, it is reasonable that a conscious or unconscious bias should exist against adopting the ideas or prototyped concepts of outsiders.

In sum, then, the incentives and organizations affecting a firm's innovation-related activities may contain many biases against—or for—outside sources of innovation data. However, once such problems become visible to the manager, many possible ways to correct them—such as specialized incentives and special interface groups—will be apparent.

Three examples of special-purpose groups a manufacturer can use to link to users with innovation data of potential interest are

Applications Laboratories. In many firms these provide free or low-cost research and development help to users interested in applying a standard product to a new application. Sometimes new product variations or markets of general interest are identified by this means.

Custom Product Groups. Such groups produce "special" products and product adaptations at customer request. In product categories where high-volume products begin as special products, these groups can be a major element in a manufacturing firm's new product R & D.

User Groups. Commonly found in the computer software area and occasionally elsewhere, user groups are a mechanism by which users of a particular type or brand of product meet and exchange ideas and information.

Note that such special-purpose interface groups are not inherently focused toward users whose needs foreshadow general demand, nor are they necessarily motivated to consider user-developed solution ideas. A firm may also orient them, for example, to specialize in helping less-skilled customers with fairly routine problems.

A firm that does wish to use such a group to link with lead users, however, may do so by using the lead user marketing research method (outlined in chapter 8) to identify the proper user subset and then realign the group to attract them and serve their needs. The lead user method has intentionally been designed to be compatible with traditional quantitative marketing research methods. Firm personnel familiar with such methods should find it reasonably easy to use. (For full methodological detail, see Urban and von Hippel.[2])

The Distributed Innovation Process as a System

To this point, I have considered how a firm might perceive and adapt to the functionally distributed innovation process as it exists in a particular innovation category. But, as I showed in chapter 7, it is also possible to modify that innovation process and the sources of innovation by shifting the profit expectations of would-be innovators. Therefore, both innovation managers and government policymakers will find it useful to understand innovation behavior at the systems level.

At this point, my research on this topic is at a very early stage. I can, however, convey the flavor and interest of the concept by means of the example of innovation in semiconductor process equipment. As we saw in chapter 2, most important innovations in semiconductor processing equipment in the period examined were typically developed by U.S. equipment users (i.e., U.S. semiconductor manufacturers) with large market shares.[3] Recently, however, discussion with industry experts indicates that many of the most important equipment innovations are being developed by user firms in Japan. The model I have developed in this book allows us to understand the causes and system-level implications of such a shift.

First, note that (as we saw in chapter 4) user firm profits from process equipment innovations will typically come on the basis of expected in-house use. Second, note that many important equipment innovations in the semiconductor field deal with increasing the density of elements on a chip. Third, note that the type of chip that is, and traditionally has been, at the leading edge of the density trend is computer memory chips (such as RAM chips). Finally, note that Japanese firms such as NEC have recently replaced U.S. firms as the major producers of such chips. It is, therefore, reasonable that Japanese user firms currently anticipate more benefit from developing this class of semiconductor process equipment innovation than U.S. firms and have therefore supplanted U.S. firms as user-innovators.

If viewed at the system level, we can also easily understand the consequences of this shift to the competitive position of U.S. and Japanese industry. When major U.S. semiconductor firms were the source of the important process equipment innovations, it improved the competitive position of those firms, but that was just the beginning of the benefits to U.S. industry. It also improved the competitive position of the U.S. equipment manufacturing firms who, owing to geographic proximity to the innovating U.S. users, were typically the first equipment builders to gain access to the innovations. Next, it improved the competitive position of noninnovating U.S. users who, being close to the source of innovation, were able to adopt more quickly than foreign rivals. Finally, it improved the competitive position of U.S. purchasers of semiconductors because they could typically gain access to the cheaper or better output resulting from application of U.S. process innovations more quickly and conveniently than could foreign buyers.

Conversely, of course, when many of the leading-edge, innovating user firms are foreign, all the interlocking effects just mentioned work to the detriment of U.S. industry. Under present conditions, U.S. semiconductor firms must often buy their equipment from foreign firms, and they still end up behind the foreign user firms that developed the technology in the first place. When U.S. equipment manufacturers lose rapid access to state-of-the-art user-developed process technology, they also suffer and decline. Finally, U.S. semiconductor buyers are forced to turn to foreign sources—often competitors—for state-of-the-art components and may lose competitive advantage thereby.

All of the elements in the examples I have just described can be seen as components in a distributed innovation process that clearly interact in a systemlike manner. Eventually, I hope we will understand such systems well enough to have a ready taxonomy of moves, countermoves, and stable states. At the moment, however, we can certainly use our present understanding to help us think about specific problems and possible solution strategies.

With respect to understanding the problem, it is clear that the system-level advantage currently held by Japanese firms in the instance of semiconductor products and equipment (formerly held by U.S. firms) is formidable in part because it is based on process innovation rather than product innovation. As we saw in chapters 4 and 5, user-developed process innovation can be effectively protected as know-how; product innovation typically cannot be and is more quickly imitated. Next, my simple economic model of the distributed innovation process allows us to understand that since key process innovations such as ion-beam implantation and X-ray lithography start out as of use to only a few advanced user firms, such user firms—*not* equipment builders—are the likely sources of the needed innovation. Next, we can understand that individual user firms in the United States are now not really in the position to make up lost ground with their own resources. This is so because firms can typically expect profits only from their own innovation-related output, and the market share of U.S. firms for the chips that require the most advanced process equipment is now low.

Our system-level understanding of the problem also allows us to think about solutions in terms of their impact on that system. We may, for example, think about lowering the innovation-related costs user-innovators expect. One approach to this would be government funding for catch-up process R & D directed to advanced U.S. user firms and targeted to leading-edge process problems. Another would involve encouraging user firms to lower R & D costs by entering into cooperative work on process development. On the benefit side of the equation, we may think about how to make innovators' expected profits from in-house process development higher or more certain. Could one, for example, make it advantageous for chip buyers to negotiate firm orders for major purchases of chips requiring new process technology far in advance? This would allow users a more certain return on their innovation investment.

Obviously, standard solutions to system-level innovation problems such as the one I have just described do not yet exist—creativity is required. At the same time I hope that I have convinced the reader that the functionally distributed innovation process is a helpful way to think about this and other innovation management problems. If so, I look forward to a bright future for related research and practitioner innovation.

Notes

1. R. Rothwell et al., "SAPPHO Updated—Project SAPPHO Phase II," *Research Policy* 3 (1974): 258–91.

2. Glen L. Urban and Eric von Hippel, "Lead User Analyses for the Development of New Industrial Products" (MIT Sloan School of Management Working Paper No. 1797–86) (Cambridge, Mass., June 1986), and *Management Science* (forthcoming).

3. All semiconductor equipment innovations sampled were developed by one of the six firms with the largest semiconductor shipments measured at the time of the innovation (Eric von Hippel, "The Dominant Role of the User in Semiconductor and Electronic Subassembly Process Innovation," *IEEE Transactions on Engineering Management* EM–24, no. 2 [May 1977]: 60–71).

REFERENCES

Achilladelis, B., A. B. Robertson, and P. Jervis. *Project SAPPHO: A Study of Success and Failure in Industrial Innovation,* 2 vols. London: Centre for the Study of Industrial Innovation, 1971.

Adamson, R. E. "Functional Fixedness as Related to Problem Solving: A Repetition of Three Experiments." *Journal of Experimental Psychology* 44 (1952): 288–91.

Adamson, R. E., and D. W. Taylor. "Functional Fixedness as Related to Elapsed Time and to Set." *Journal of Experimental Psychology* 47 (1954): 122–26.

Allen, Robert C. "Collective Invention." *Journal of Economic Behavior and Organization* 4, no. 1 (March 1983): 1–24.

Allen, Thomas J. *Managing the Flow of Technology.* Cambridge, Mass.: MIT Press, 1977.

———. "Studies of the Problem-Solving Process in Engineering Design." *IEEE Transactions on Engineering Management* EM–13, no. 2 (June 1966): 72–83.

Allen, Thomas J., Diane B. Hyman, and David L. Pinckney. "Transferring Technology to the Small Manufacturing Firm: A Study of Technology Transfer in Three Countries." *Research Policy* 12, no. 4 (August 1983): 199–211.

Allen, T. J., and D. G. Marquis. "Positive and Negative Biasing Sets: The Effects of Prior Experience on Research Performance." *IEEE Transactions on Engineering Management* EM–11, no. 4 (December 1964): 158–61.

Anderson, W. A., and J. T. Arnold. "A Line-Narrowing Experiment." *Physical Review* 94, no. 2 (15 April 1954): 497–98.

Arrow, Kenneth J. "The Economic Implications of Learning By Doing." *Review of Economic Studies* 29 (June 1962): 155–73.

———. "Economic Welfare and the Allocation of Resources for Invention." In *The Rate and Direction of Inventive Activity: Economic and Social Factors,* A Report of the National Bureau of Economic Research, 609–25. Princeton, N.J.: Princeton University Press, 1962.

Axelrod, Robert. *The Evolution of Cooperation.* New York: Basic Books, 1984.

Barnett, Donald F., and Louis Schorsch. *Steel: Upheaval in a Basic Industry.* Cambridge, Mass.: Ballinger, 1983.

Berger, Alan J. "Factors Influencing the Locus of Innovation Activity Leading to Scientific Instrument and Plastics Innovation." SM thesis, Sloan School of Management, MIT, Cambridge, Mass., 1975.

"The Big Deal McDonnell Douglas Turned Down: Delta's $3 Billion Order for 60 Jets Went to Boeing." *Business Week* 2665 (1 December 1980): 81–82.

Birch, H. G., and H. J. Rabinowitz. "The Negative Effect of Previous Experience on Productive Thinking." *Journal of Experimental Psychology* 41 (1951): 121–26.

Blalock, Hubert M. *Social Statistics*. Rev. 2nd ed. New York: McGraw–Hill, 1979.

Bloch, F. "Line-Narrowing by Macroscopic Motion." *Physical Review* 94, no. 2 (15 April 1954): 496–97.

Boyden, Julian W. "A Study of the Innovative Process in the Plastics Additives Industry." SM thesis, Sloan School of Management, MIT, Cambridge, Mass., 1976.

Buer, Terje Kristian. "Investigation of Consistent Make or Buy Patterns of Selected Process Machinery in Selected U.S. Manufacturing Industries." PhD diss., Sloan School of Management, MIT, Cambridge, Mass., 1982.

"Chemicals & Additives '83: A *Modern Plastics* Special Report." *Modern Plastics* 60, no. 9 (September 1983): 53–76.

Clarke, Roderick W. "Innovation in Liquid Propellant Rocket Technology." PhD diss., Stanford University, Stanford, Calif., 1968.

Collins, H. M. "Tacit Knowledge and Scientific Networks." In *Science in Context: Readings in the Sociology of Science,* ed. Barry Barnes and David Edge, 44–64. Cambridge, Mass.: MIT Press, 1982.

Commerce Clearing House. "National Cooperative Research Act of 1984 (Act of October 11, 1984, Public Law 98–462)." *Trade Regulation Reports,* pp. 27,082–086.

"Construction Equipment: Ten Years of Change." *Engineering News Record* 21 (February 1963): 45.

Corey, E. Raymond. *The Development of Markets for New Materials: A Study of Building New End-Product Markets for Aluminum, Fibrous Glass, and the Plastics.* Boston: Division of Research, Graduate School of Business Administration, Harvard University, 1956.

Council of Economic Advisers Annual Report (1981). In *Economic Report of the President, Transmitted to the Congress, January 1981, Together with the Annual Report of the Council of Economic Advisers,* 21–213. Washington, D.C.: U.S. Government Printing Office, 1981.

Duncker, Karl. "On Problem Solving," trans. Lynne S. Lees. *Psychological Monographs* 58, no. 5 (1945).

Enos, John Lawrence. *Petroleum Progress and Profits: A History of Process Innovation.* Cambridge, Mass.: MIT Press, 1962.

Federico, P. J. *Distribution of Patents Issued to Corporations* (*1939–44*). Study No. 3 of the Subcommittee on Patents, Trademarks, and Copyrights. Washington, D.C.: U.S. Government Printing Office, 1957.

Foxall, Gordon R., Francis S. Murphy, and Janet D. Tierney. "Market Development in Practice: A Case Study of User-Initiated Product Innovation." *Journal of Marketing Management* 1 (1985): 201–11.

Freeman, Christopher. *The Economics of Industrial Innovation.* Harmondsworth, Eng.: Penguin Books, 1974.

———. "Research and Development in Electronic Capital Goods." *National Institute Economic Review* 34 (November 1965): 40–91.

Gaskins, Darius W., Jr. "Dynamic Limit Pricing: Optimal Pricing Under Threat of Entry." *Journal of Economic Theory* 3 (September 1971): 306–22.

Golding, A. M. "The Semiconductor Industry in Britain and the United States: A Case Study in Innovation, Growth and the Diffusion of Technology." PhD diss., University of Sussex, Eng., 1971.

Griliches, Zvi, ed. *R & D, Patents and Productivity.* Chicago: University of Chicago Press, 1987.

Griliches, Zvi, and Jacob Schmookler. "Comment: Inventing and Maximizing." *American Economic Review* 53, no. 10 (September 1963): 725–29.

Hayes, Robert H., and Steven C. Wheelwright. *Restoring Our Competitive Edge: Competing Through Manufacturing*. New York: Wiley, 1984.

Hollander, Samuel. *The Sources of Increased Efficiency: A Study of Du Pont Rayon Plants*. Cambridge, Mass.: MIT Press, 1965.

"How 'Silicon Spies' Get Away with Copying." *Business Week* 2633 (21 April 1980): 178, 182, 187–88.

IMS America. *Semi-Annual Audit of Laboratory Tests, Hospital Labs*. January–June 1977; July–December 1977. Ambler, Penn.: IMS America, n.d.

Institute for Defense Analyses. *The Effects of Patent and Antitrust Laws, Regulations, and Practices on Innovation*. 3 vols. Arlington, Va.: National Technical Information Service, 1976.

Isenson, Raymond S. "Project Hindsight: An Empirical Study of the Sources of Ideas Utilized in Operational Weapon Systems." In *Factors in the Transfer of Technology,* ed. William H. Gruber and Donald G. Marquis, 155–76. Cambridge, Mass.: MIT Press, 1969.

Jewkes, John, David Sawers, and Richard Stillerman. *The Sources of Invention,* 2nd ed. New York: Norton, 1969.

Johnson, P. S. *Co-operative Research in Industry: An Economic Study*. New York: Wiley, 1973.

Juhasz, Andrew Anthony, Jr. "The Pattern of Innovation Exhibited in the Development of the Tractor Shovel." SM thesis, Sloan School of Management, MIT, Cambridge, Mass., 1975.

Kamien, Morton I., and Nancy L. Schwartz. "Market Structure and Innovation: A Survey." *Journal of Economic Literature* 13, no. 1 (March 1975): 1–37.

Kantrow, Alan M. "The Strategy-Technology Connection." In *The Management of Technological Innovation,* 3–9. Cambridge, Mass.: Harvard Business Review, 1982.

Kingston, William. *Innovation: The Creative Impulse in Human Progress*. London: John Calder, 1977.

Kitti, Carole. "Patent Invalidity Studies: A Survey." National Science Foundation, Division of Policy Research and Analysis, Washington, D.C., January 1976.

Klein, Bernard, James H. Kaufman, and Stanley Morgenstern. "Determination of Serum Calcium by Automated Atomic Absorption Spectroscopy." In *Automation in Analytical Chemistry: Technicon Symposia*. Vol. 1, 10–14. New York: Mediad, 1967.

Knight, Kenneth E. "A Study of Technological Innovation: The Evolution of Digital Computers." PhD diss., Carnegie Institute of Technology, Pittsburgh, Penn., 1963.

Kodama, Fumio. "Technological Diversification of Japanese Industry." *Science* 233, no.4761 (18 July 1986): 291–304.

Kuznets, Simon. "Inventive Activity: Problems of Definition and Measurement." In *The Rate and Direction of Inventive Activity: Economic and Social Factors,* A Report of the National Bureau of Economic Research, 19–43. Princeton, N.J.: Princeton University Press, 1962.

Ladd, Hon. David, Commissioner of Patents, Statement Before the Patents, Trademarks, and Copyrights Subcommittee of the Judiciary Committee, U.S. Senate, September 4, 1962, re: S.2225. Quoted in Elmer J. Gorn, *Economic Value of Patents, Practice and Invention Management*. New York: Reinhold, 1964.

Lehmann, Walter G. "Innovation in Electron Microscopes and Accessories." SM thesis, Sloan School of Management, MIT, Cambridge, Mass., 1975.

Levin, Richard C. "A New Look at the Patent System." *American Economic Review* 76, no. 2 (1 May 1986): 199–202.

———. "Technical Change, Barriers to Entry, and Market Structure." *Economica* 45 (November 1978): 347–61.

Levin, Richard C., and Richard R. Nelson. "Survey Research on R & D Appropriability and Technological Opportunity. Part I: Appropriability." Working Paper, Yale University. New Haven, Conn., July 1984.

Levitt, Theodore. *The Marketing Imagination.* New York: Free Press, 1983.

Lionetta, William G., Jr. "Sources of Innovation Within the Pultrusion Industry." SM thesis, Sloan School of Management, MIT, Cambridge, Mass., 1977.

Luchins, A. S. "Mechanization in Problem-Solving: The Effect of *Einstellung*." *Psychological Monographs* 54 (1942).

Mahon, J. "Trade Secrets and Patents Compared." *Journal of the Patent Office Society* 50, no. 8 (August 1968): 536.

Mansfield, Edwin. *The Economics of Technological Change.* New York: Norton, 1968.

———. "How Rapidly Does New Industrial Technology Leak Out?" *Journal of Industrial Economics* 34, no. 2 (December 1985): 217–23.

———. *Industrial Research and Technological Innovation: An Econometric Analysis.* New York: Norton, 1968.

Mansfield, Edwin, et al. "Social and Private Rates of Return from Industrial Innovations." *Quarterly Journal of Economics* 91, no. 2 (May 1977): 221–40.

Marples, David L. "The Decisions of Engineering Design." *IRE Transactions on Engineering Management,* June 1961: 55.

Marx, Karl. *Capital: A Critique of Political Economy* (1867). Vol. 1 (Modern Library Edition, ed. Ernest Untermann). New York: Random House, 1936.

Materials Advisory Board, Division of Engineering, National Research Council. *Report of the Ad Hoc Committee on Principles of Research-Engineering Interaction.* Publication MAB–222–M. Washington, D.C.: National Academy of Sciences-National Research Council, July 1966.

Mauss, Marcel. *The Gift: Forms and Functions of Exchange in Archaic Societies,* trans. Ian Cunnison. Quotation from *The Havamal, with Selections from other Poems in the Edda,* xiv. Glencoe, Ill.: Free Press, 1954.

Dennis L. Meadows, "Accuracy of Technical Estimates in Industrial Research Planning: Data Appendix." MIT Sloan School of Management Working Paper No. 301–67. Cambridge, Mass., December 1967.

———. "Estimate Accuracy and Project Selection Models in Industrial Research." *Industrial Management Review* 9, no. 3 (Spring 1968): 105–19.

Merton, Robert K. *The Sociology of Science: Theoretical and Empirical Investigations,* ed. Norman W. Storer. Chicago: University of Chicago Press, 1973.

Myers, Sumner, and Donald G. Marquis. *Successful Industrial Innovations: A Study of Factors Underlying Innovation in Selected Firms.* Washington, D.C.: National Science Foundation, NSF 69–17, 1969.

National Academy of Sciences. *Applied Science and Technological Progress,* A Report to the Committee on Science and Astronautics, U.S. House of Representatives, GP–67–0399. Washington, D.C.: U.S. Government Printing Office, June 1967.

National Library of Medicine. *Medical Subject Headings—Annotated Alphabetical*

List, 1978. Springfield, Mass.: U.S. Department of Commerce, National Technical Information Service (No. PB–270–894), 1978.
National Research Council of the National Academy of Sciences. *Chemistry: Opportunities and Needs*. Washington, D.C.: National Academy of Sciences, 1965.
Nelson, Richard R. "The Role of Knowledge in R & D Efficiency." *Quarterly Journal of Economics* 97, no. 3 (August 1982): 453–70.
———. "The Simple Economics of Basic Scientific Research." *Journal of Political Economy* 67, no. 3 (June 1959): 297–306.
Nelson, Richard R., and Sidney G. Winter. *An Evolutionary Theory of Economic Change*. Cambridge, Mass.: Harvard University Press, 1982.
Nemeth, Edward L. "Mini-Midi Mills—U.S., Canada, and Mexico." *Iron and Steel Engineer* 61, no. 6 (June 1984): 25–55.
Newell, Allen, and Robert F. Sproul. "Computer Networks: Prospects for Scientists." *Science* 215, no. 4534 (12 February 1982): 843–52.
Olson, Mancur. *The Logic of Collective Action: Public Goods and the Theory of Groups*. Cambridge, Mass.: Harvard University Press, 1965.
Pakes, Ariel, and Mark Schankerman. "An Exploration into the Determinants of Research Intensity." In *R & D, Patents and Productivity*, ed. Zvi Griliches. Chicago: University of Chicago Press, 1987.
Park, C. Whan, and V. Parker Lessig. "Familiarity and Its Impact on Consumer Decision Biases and Heuristics." *Journal of Consumer Research* 8, no. 2 (September 1981): 223–30.
Pavitt, Keith. "Sectoral Patterns of Technical Change: Towards a Taxonomy and a Theory." *Research Policy* 13, no. 6 (December 1984): 343–73.
Peck, Merton J. "Inventions in the Postwar American Aluminum Industry." In *The Rate and Direction of Inventive Activity: Economic and Social Factors*, A Report of the National Bureau of Economic Research, 279–98. Princeton, N.J.: Princeton University Press, 1962.
Perry, B. W., et al. *A Field Evaluation of the Du Pont Automatic Clinical Analyzer*. Wilmington, Del.: Du Pont, n.d.; 2nd printing, January 1978.
Pickens, Stephen H. "Pultrusion—The Accent on the Long Pull." *Plastics Engineering* 31, no. 7 (July 1975): 16–21.
Pratten, C. F. *Economies of Scale in Manufacturing Industry*. Cambridge: Cambridge University Press, 1971.
Rauch, James A., ed. *The Kline Guide to the Plastics Industry*. Fairfield, N.J.: Charles H. Kline, 1978.
Rautela, Gopal S., and Raymond J. Liedtke. "Automated Enzymic Measurement of Total Cholesterol in Serum." *Clinical Chemistry* 24, no. 1 (January 1978): 108–14.
Regan, Dennis T., and Russell Fazio. "On the Consistency Between Attitudes and Behavior: Look to the Method of Attitude Formation." *Journal of Experimental Social Psychology* 13, no. 1 (January 1977): 28–45.
Rice, Ronald E., and Everett M. Rogers. "Reinvention in the Innovation Process." *Knowledge: Creation, Diffusion, Utilization* 1, no. 4 (June 1980): 499–514.
Roberts, Edward B. "Exploratory and Normative Technological Forecasting: A Critical Appraisal." *Technological Forecasting* 1 (1969): 113–27.
———. "A Simple Model of R & D Project Dynamics." *R & D Management* 5, no. 1 (October 1974): 1–15.
Roberts, John H., and Glen L. Urban. *New Consumer Durable Brand Choice: Model-*

ing Multiattribute Utility, Risk, and Dynamics. MIT Sloan School of Management Working Paper No. 1636–85. Cambridge, Mass., 1985.

Rogers, Everett M., J. D. Eveland, and Constance A. Klepper. "The Diffusion and Adoption of GBF/DIME Among Regional and Local Governments." Working Paper, Institute for Communications Research, Stanford University, Stanford, Calif.; paper presented at the Urban and Regional Information Systems Association, Atlanta, Ga., 31 August 1976.

Rogers, Everett M., with F. Floyd Shoemaker. *Communication of Innovations: A Cross-Cultural Approach, 2nd ed.* New York: Free Press, 1971.

Rosenberg, Nathan. *Perspectives on Technology.* Cambridge: Cambridge University Press, 1976.

———. "Science, Invention and Economic Growth." *Economic Journal* 84, no. 333 (March 1974): 90–108.

———. ed. *Research on Technological Innovation Management and Policy.* Vol. 1. Greenwich, Conn.: JAI Press, 1983.

Rosenbloom, Richard S., "Technological Innovation in Firms and Industries: An Assessment of the State of the Art." In *Technological Innovation: A Critical Review of Current Knowledge,* ed. P. Kelly and Marvin Kranzberg. San Francisco: San Francisco Press, 1978.

Rothwell, R., et al. "SAPPHO Updated—Project SAPPHO Phase II." *Research Policy* 3 (1974): 258–91.

Rothwell, Roy, and Walter Zegveld. *Innovation and the Small and Medium Sized Firm.* Hingham, Mass.: Kluwer–Nijhoff, 1981.

Schelling, Thomas C. *Choice and Consequence.* Cambridge, Mass.: Harvard University Press, 1984.

Scherer, F. M. *Innovation and Growth: Schumpeterian Perspectives.* Cambridge, Mass.: MIT Press, 1984.

Scherer, F. M., et al. *Patents and the Corporation.* 2nd ed. Boston: James Galvin and Associates, 1959.

Schmookler, Jacob. "Changes in Industry and in the State of Knowledge as Determinants of Industrial Invention." In *The Rate and Direction of Inventive Activity: Economic and Social Factors,* A Report of the National Bureau of Economic Research, 195–232. Princeton, N.J.: Princeton University Press, 1962.

———. *Invention and Economic Growth.* Cambridge, Mass.: Harvard University Press, 1966.

Schon, Donald A. *Technology and Change.* New York: Delacorte Press, 1967.

Schumpeter, Joseph A. *Capitalism, Socialism and Democracy.* 3rd ed. New York: Harper & Row, 1950.

Shapley, Deborah. "Electronics Industry Takes to 'Potting' Its Products for Market." *Science* 202, no. 4370 (24 November 1978): 848–49.

Shaw, Brian. "Appropriation and Transfer of Innovation Benefit in the U.K. Medical Equipment Industry." *Technovation* 4, no. 1 (February 1986): 45–65.

———. "The Role of the Interaction Between the User and the Manufacturer in Medical Equipment Innovation." *R & D Management* 15, no. 4 (October 1985): 283–92.

Shimshoni, Daniel. "Aspects of Scientific Entrepreneurship." PhD diss., Harvard University, Cambridge, Mass., 1966.

Shocker, Allan D., and V. Srinivasan. "Multiattribute Approaches for Product Concept Evaluation and Generation: A Critical Review." *Journal of Marketing Research* 16, no. 2 (May 1979): 159–80.

Silk, Alvin J., and Glen L. Urban. "Pre-Test-Market Evaluation of New Packaged Goods: A Model and Measurement Methodology." *Journal of Marketing Research* 15, no. 2 (May 1978): 171–91.

Sittig, Marshall. *PolyAcetyl Resins*. Houston, Tex.: Gulf Publishing Company, 1963.

Skinner, Wickam, and David C. D. Rogers. *Manufacturing Policy in the Electronics Industry: A Casebook of Major Production Problems*. 3rd ed. Homewood, Ill.: Richard D. Irwin, 1968.

Smith, Adam. *An Inquiry into the Nature and Causes of the Wealth of Nations* (1776; 5th ed., 1789). (Modern Library Edition, ed. Edwin Cannan.) New York: Random House, 1937.

Smithson, L. H. *Overview of the Clinical Laboratory Market*. Menlo Park, Calif.: Stanford Research Institute, n.d.

Solomon, Stephen D. "MetPath's Biggest Test." *Fortune* 97, no. 6 (27 March 1978): 114–16, 118.

Solow, Herbert. "Delrin: Du Pont's Challenge to Metals." *Fortune* 60, no. 2 (August 1959): 116–19.

Solow, Robert M. "Technical Change and the Aggregate Production Function." *Review of Economics and Statistics* 39, no. 3 (August 1957): 312–20.

Spital, Francis Clifford. "The Role of the Manufacturer in the Innovation Process for Analytical Instruments." PhD diss., Sloan School of Management, MIT, Cambridge, Mass., 1978.

Stigler, George J. *The Organization of Industry*. Homewood, Ill.: Richard D. Irwin, 1968.

Taylor, C. T., and Z. A. Silberston. *The Economic Impact of the Patent System: A Study of the British Experience*. Cambridge: Cambridge University Press, 1973.

Tilton, John E. *International Diffusion of Technology: The Case of Semiconductors*. Washington, D.C.: Brookings Institution, 1971.

Torgerson, Warren S. *Theory and Methods of Scaling*. New York: Wiley, 1958.

U.S. Department of Commerce, Bureau of the Census. *Current Industrial Reports: Selected Instruments and Related Products*. MA38–B (80)–1 January 1982, SIC Code 38112. Washington, D.C.: U.S. Government Printing Office, 1982.

———. *1977 Census of Manufactures,* Vol. 2, *Industry Statistics*. Washington, D.C.: U.S. Government Printing Office, 1980.

Urban, Glen L., and John R. Hauser. *Design and Marketing of New Products*. Englewood Cliffs, N.J.: Prentice–Hall, 1980.

Urban, Glen L., and Eric von Hippel. "Lead User Analyses for the Development of New Industrial Products." MIT Sloan School of Management Working Paper No. 1797–86. Cambridge, Mass., June 1986. Forthcoming in *Management Science*.

Urban, Glen L., et al. "Market Share Rewards to Pioneering Brands: An Empirical Analysis and Strategic Implications." *Management Science* 32, no. 6 (June 1986): 645–59.

Utterback, James M. "The Process of Innovation: A Study of the Origination and Development of Ideas for New Scientific Instruments." *IEEE Transactions on Engineering Management* EM–18, no. 4 (November 1971): 124–31.

Utterback, James M., and William J. Abernathy. "A Dynamic Model of Process and Product Innovation." *Omega, The International Journal of Management Science* 3, no. 6 (1975): 639–56.

VanderWerf, Pieter A. "Explaining the Occurrence of Industrial Process Innovation by Materials Suppliers with the Economic Benefits Realizable from Innovation." PhD diss. Sloan School of Management, MIT, Cambridge, Mass., 1984.

———. “Parts Suppliers as Innovators in Wire Termination Equipment.” MIT Sloan School of Management Working Paper No. 1289–82. Cambridge, Mass., March 1982. Forthcoming in *Research Policy* entitled “Supplier Innovation in Electronic Wire and Cable Preparation Equipment.”

von Hippel, Eric. “Appropriability of Innovation Benefit as a Predictor of the Source of Innovation.” *Research Policy* 11, no. 2 (April 1982): 95–115.

———. “Cooperation Between Competing Firms: Informal Know-How Trading.” MIT Sloan School of Management Working Paper No. 1759–86. Cambridge, Mass., March 1986. Forthcoming in *Research Policy.*

———. “A Customer-Active Paradigm for Industrial Product Idea Generation.” *Research Policy* 7, no. 3 (July 1978): 240–66.

———. “The Dominant Role of Users in the Scientific Instrument Innovation Process.” *Research Policy* 5, no. 3 (July 1976): 212–39.

———. “The Dominant Role of the User in Semiconductor and Electronic Subassembly Process Innovation.” *IEEE Transactions on Engineering Management* EM–24, no. 2 (May 1977): 60–71.

———. “Lead Users: A Source of Novel Product Concepts.” *Management Science* 32, no. 7 (July 1986): 791–805.

———. “Learning from Lead Users.” In *Marketing in an Electronic Age,* ed. Robert D. Buzzell, 308–17. Cambridge, Mass.: Harvard Business School Press, 1985.

———. “Successful Industrial Products from Customer Ideas.” *Journal of Marketing* 42, no. 1 (January 1978): 39–49.

———. “Testing the Correlation Between the Functional Locus of Innovation and Appropriable Innovation Benefit.” MIT Sloan School of Management Working Paper No. 1688–85. Cambridge, Mass., August 1985.

———. “Transferring Process Equipment Innovations from User-Innovators to Equipment Manufacturing Firms.” *R & D Management* 8, no. 1 (October 1977): 13–22.

———. “Users as Innovators.” *Technology Review* 80, no. 3 (January 1978): 3–11.

———. “The User's Role in Industrial Innovation.” *TIMS Studies in the Management Sciences* 15 (1980): 53–65.

von Hippel, Eric, and Stan N. Finkelstein. “Analysis of Innovation in Automated Clinical Chemistry Analyzers.” *Science & Public Policy* 6, no. 1 (February 1979): 24–37.

Williamson, Oliver E. “Innovation and Market Structure.” *Journal of Political Economy* 73, no. 1 (February 1965): 67–73.

Wilson, R. W. “The Sale of Technology Through Licensing.” PhD diss., Yale University, New Haven, Conn., 1975.

Yorsz, Walter. “A Study of the Innovative Process in the Semiconductor Industry.” SM thesis, Sloan School of Management, MIT, Cambridge, Mass., 1976.

APPENDIX

Innovation Histories

INTRODUCTION

In the course of collecting and cross-checking the data on the functional locus of innovation coded in chapters 2 and 3 of this book, my students and I generated what were, in effect, minihistories of the innovations under study. I present these as a data base for others interested in exploring the functional source of innovation. Also, I will mention here a few hard-won bits of practical wisdom regarding data collection on this topic that others may find useful.

Suggestions Regarding Data Collection

Our data collection strategy was built around the principle of *independent* access to the various functional categories of firms and individuals that may have played a role in an innovation under study. That is, we searched all likely functional communities directly for evidence of possible involvement in the innovation process. This strategy proved very useful. Since "success has many fathers," reliance on a single data source as a gateway to others would almost certainly have introduced a systematic bias to our data. For example, a data collection strategy that relied on manufacturers to point out the innovation-related contributions of users and others would predictably produce findings biased toward the contributions of manufacturers.

A short anecdote may serve to give the reader the flavor of the problem. We inquired of the first responsible project engineer at what was apparently the first commercial manufacturer of an instrument as to its innovation history. "All mine," he said. When we ran a computer search of the chemical literature, we found three articles, published several years earlier, describing experimental versions of the innovation and the interesting results obtained through their use. A parallel search of trade journal ads produced information about a functionally equivalent Canadian instrument that had apparently been introduced to the marketplace prior to the introduction of the U.S. company's version.

We went back to our original informant to discuss these findings. He admitted, in honest confusion, that he was aware of the articles and product we had found, but he had not mentioned them to us because he did not consider them to be related to his innovation. To be sure, the function and operating principles of the experimental devices and the Canadian instrument were the same, but the *product engineering* of his device was entirely his own—and this had been the innovation, in his view.

Our data collection procedure used three major elements for each sampled innovation involved. First, we started our work by identifying the first firm to develop and commercialize an innovation and the date of commercial introduction. This was done by asking personnel of firms now manufacturing "me-too" devices if they knew which was the first firm to commercialize them *and* by asking expert users, manufacturers, and suppliers of the innovation. Ambiguities were cleared up by searching the technical or trade literature and seeking the earliest references to, advertisements regarding, and/or new product announcements for the innovation in question.

When the first commercializing firm and the date of commercialization were firmly established, we interviewed, either in person or by telephone, everyone at the commercializing firm who claimed substantial, firsthand knowledge of the innovation work. As a routine part of a structured interview, each interviewee was asked to provide the names of others he or she felt might have some important information to contribute, and these individuals were contacted in turn.

Second, in addition to our interviews with personnel in the first commercializing firms, we searched the technical and trade literature of the period *prior* to the first commercial innovation, seeking references to products or processes functionally similar to the innovation being studied. Authors of relevant articles were contacted and interviewed, usually by telephone. As part of the interview, they were asked for the names of knowledgeable people in user and other firms, and these people were subsequently contacted.

Where we could identify very early purchasers of the commercialized innovation, we also canvassed these firms for information regarding their contribution, if any, to the innovation and/or the name of individuals who might have information bearing on the innovation process, and so forth. Thus, insofar as possible, we interviewed all key individuals identified as being directly involved with each innovation studied.

Third, information from the various sources was assembled, discrepancies were noted, and interviewees with information bearing on the discrepancies were contacted again for further discussion. Some areas of confusion were cleared up by means of this process; others were not. We always attempted to accurately preserve differing versions of events where they existed and did not attempt to determine "who was right." If proper coding of an item would require us to make such a judgment, we coded it NA (not available).

Other Matters

Our ability to accurately collect data regarding innovations that had often been developed a number of years ago was greatly enhanced by our definition of an innovator (discussed earlier) as one who had developed an innovative product or process to a concrete, usable state. Development work carried to this stage usually leaves many contemporaneously generated documents and artifacts—reports, publications, prototypes—that are most helpful in reconstructing innovation process events.

Defining the innovator as the first to develop a product or process to a useful state ignores any contributions by firms or individuals that might be valuable but that do not reach the state of implementation. This is a loss, but I am not sure how much of a loss. As has been shown by studies such as Project Hindsight,[1] any innovation is built upon a great web of earlier developments in technology and science stretching back, certainly, to the Renaissance and even earlier. Is it meaningful to try and allocate the locus of a particular innovation among all these precursors? For our purposes I think not. We are interested in determining who builds upon prior art and science to produce very specific innovations. In practice I find that at this level of specificity the people who innovate have a particular need or market in mind that can only be served by a completed device. Since there is little benefit to be derived from half an innovation, I find, as a practical matter, that innovators strive to bring the innovations they work on to a useful state. Therefore, my definition of *the* innovator as the firm or individual that first develops a specific product or process to a completed, useful state appears to accurately reflect the realities of this type of innovation.

Finally, in this book I focus on the three innovation categories of user, supplier, and manufacturer. For other research purposes, one might wish to segment such major categories more finely. For example, one could divide the general category of users of scientific instruments into subsets such as research users, teachers, and pupils—each of which use a scientific instrument in different ways. Such subsets will cause no trouble in analyses of the functional locus of innovation as long as they are made clear. For example, if the distinction between research/user of scientific instruments and teacher/ user is understood, one can separately analyze the innovation-related benefits attendant upon firms holding each of these types of functional relationship and determine the appropriate role of each in the innovation process. (Appropriate segmentation of user subsets can sometimes be aided by application of segmentation techniques often used by marketing researchers.[2])

Notes

1. Raymond S. Isenson, "Project Hindsight: An Empirical Study of the Sources of Ideas Utilized in Operational Weapon Systems," in *Factors in the Transfer of Technology,* ed. William H. Gruber and Donald G. Marquis (Cambridge, Mass.: MIT Press, 1969), 155–76.
2. Glen L. Urban and John R. Hauser, *Design and Marketing of New Products* (Englewood Cliffs, N.J.: Prentice-Hall, 1980).

DATA SET FOR SCIENTIFIC INSTRUMENT INNOVATIONS

The data set for scientific instrument innovations contains information on the gas chromatograph (GC), the nuclear magnetic resonance (NMR) spectrometer, and the transmission electron microscope (TEM). In the following pages the innovation history of each basic instrument type is followed by histories of the major improvements to it that have been commercialized over the years. The histories of all the innovations not coded NA in chapter 2 will be found in this data set, with two exceptions: the innovations identified in our study of ultraviolet absorption spectrophotometry and all minor improvements to electron microscopy. In these instances we have no data beyond that presented in chapter 2.

Selection criteria for the samples in this data set will be found in chapter 2.

THE GAS CHROMATOGRAPH (GC)

The gas chromatograph is used for the quantitative and qualitative analysis of unknown chemical mixtures. It provides much faster and more accurate analyses than earlier wet chemistry methods and is very frequently used.[1] The GC operates by physically separating a chemical mixture into its components. The mixture is passed in the form of a gas over a surface containing a partitioning agent that selectively *ad*sorbs its components, thus slowing the rate of movement of some components relative to others. The adsorbing surface is contained in a column. Gas injected into one end of the column as a mixture (combined with carrier gas) emerges from the other end as a sequence of components that pass in series through any of several types of detectors for identification.

Development History

The analytical technique embodied in the gas chromatograph—gas–liquid partition chromatography—can be traced back to a 1941 paper by A.J.P. Martin and R.L.M. Synge.[2] This paper suggested the idea and described the process theoretically—an accomplishment for which the authors were later awarded a Nobel Prize. In 1952 Martin suggested to A. T. James—a young scientist working with him at the Mill Hill Medical Research Laboratories in England—that he try to build a GC along the lines outlined in the 1941 paper. James did, and the device worked. Their initial paper in 1952 described the apparatus and gave some results they had obtained with it.[3]

After the publication of the James and Martin article, many scientists in industry and universities began to experiment with the technique. By 1953–54 there were perhaps two dozen homemade GC devices in use around the world.[4]

Commercialization

Commercial GCs were first produced in 1954–55. British instrument firms (among them Griffin and George, London, and Metropolitan Vickers Electrical Company, Manchester) began producing commercial instruments before American firms entered the market. The first two American firms to begin commercial production in the spring of 1955 were the Burrell Corporation and the Perkin–Elmer Corporation.[5] In the instance of Perkin–Elmer, we have learned by interview that the transfer to commercialization came about as follows: Dr. Z. V. Williams, vice president of Perkin–Elmer, often traveled to England on company business and had contact among scientists there (among other products, Perkin–Elmer sold spectrophotometers to industrial and university scientists). On one of these trips in 1953, Williams heard of gas–liquid partition chromatography and suggested to Harry Hausdorff, a young employee of Perkin–Elmer with a background in chemistry, that it might be worth looking into as a commercial possibility. Hausdorff made a trip to England, visited laboratories where homebuilt gas chromatography apparatus was in use, attended a lecture at Oxford on gas chromatography, and came back (he recalls) with about 20 journal articles on the subject. Dr. Hausdorff was convinced of the commercial potential of gas chromatography after his trip but had some difficulty interesting his superiors in the project (they noted that the device had no optical parts and optics was, after all, Perkin–Elmer's forte). Eventually, Hausdorff prevailed and was allowed to proceed with commercialization.[6]

Notes

1. National Research Council of the National Academy of Sciences, *Chemistry: Opportunities and Needs* (Washington, D.C.: National Academy of Sciences, 1965).
2. A.J.P. Martin and R.L.M. Synge, "A New Form of Chromatogram Employing Two Liquid Phases ," *Biochemical Journal* 35 (1941): 1358–68.
3. A. T. James. and A.J.P. Martin, "Gas–liquid Partition Chromatography: The

Separation and Micro-estimation of Volatile Fatty Acids from Formic Acid to Dodecanoic Acid," *Biochemical Journal* 50 (1952): 679–90.
4. L. J. Ettre, personal communication while at Perkin–Elmer Corp., Norwalk, Conn.
5. L. J. Ettre, "The Development of Gas Chromatography," *Journal of Chromatography* 112 (1975): 1–26.
6. Dr. H. H. Hausdorff, interview at Perkin–Elmer Corp., Norwalk, Conn., 1971.

Temperature Programming

Raising the temperature of the gas chromatograph allows one to analyze substances that volatilize at greater-than-room temperatures. Raising the temperature of the instrument during an analysis in a preprogrammed manner allows one to rapidly analyze a sample containing components of very different boiling points. This very useful improvement in performance was obtained by modifying the various components of the GC to tolerate high temperatures and then installing the entire instrument in a special-purpose temperature-controlled oven.

Development History

Courtney Phillips appears to have originated temperature programming at Oxford University in 1952.[1] Gas chromatographs with temperature programming were then built by numerous users for their own use. Over 20 publications related to temperature programming appeared in the literature prior to the appearance of the first commercial instrument on the marketplace.[2]

Commercialization

Frank Martinez was a glass blower at Du Pont who was involved in constructing several of the temperature-programmed GCs built at Du Pont for the use of that firm's scientists. When interest was shown in the device by non–Du Pont people, Martinez left the firm and formed F&M Scientific Corporation in April 1959 to manufacture a Du Pont-designed instrument that included temperature programming—reportedly a design developed by S. Dal Nogare, a Du Pont scientist. This new firm was the first to offer temperature-controlled GCs for commercial sale.

Notes

1. J. Griffiths, D. James, and C. Phillips, "Gas Chromatography," *Analyst* 77 (December 1952): 897–904.
2. W. E. Harris and H. W. Habgood, "Annotated Bibliography of Programmed Temperature Gas Chromatography: 1952–1964, Part I. 1952–1961," *Journal of Gas Chromatography* 4 (April 1966): 144–46.

Capillary Column

The capillary column is a long thin tube coated on the inside with partitioning agent. When the short, granule-packed column normally used in a GC is replaced by such a column, the sensitivity and resolution of the instrument is significantly improved.

To understand the reason for this improvement I must elaborate slightly on my previous description of the operating principles of a GC. As noted earlier, GCs separate the components of a gaseous chemical mixture by passing them through a granule-packed column. A coating on the granules adsorbs (clings to) some components of the mixture more strongly than others. The more strongly a particular component is adsorbed, the more slowly it is moved through the column by the stream of carrier gas passing steadily through the column. The result of this effect is the physical separation of the components of the mixture, with those least strongly adsorbed emerging from the end of the column first.

Efficiency of component separation attainable by the GC is clearly lessened because the gas must pass through a column of granules: Gas flow will inevitably be less rapid between some granules than others, and some paths available to gas molecules through the maze of granules will be longer than others. The capillary column, being a simple hollow cylinder coated with partitioning agent, greatly reduces these sources of error.

Development History

The capillary column concept was developed by Marcel Golay, a mathematician who worked as a consultant to Perkin–Elmer Corporation. According to Harry Hausdorff,[1] the gas chromatography group at Perkin–Elmer set Golay the task of developing a "universal" column. Their motivation was to reduce the number of types of specialized prepared granular columns they had to keep in inventory to service customer needs. Golay presented the concept of the capillary column in 1957.[2] In 1958 both he[3] and a user group[4] presented practical results obtained with such columns at a conference in Amsterdam.

Commercialization

Perkin–Elmer, the employer of Golay, observed the strong user interest displayed in capillary columns and quickly commercialized them, first making them available in March 1959.

Because Perkin–Elmer was the first to develop the capillary column as well as the first to commercialize it, the firm was in a position to patent the innovation and did so. Interestingly, however, tentative moves to enforce the patent were abandoned in the face of user protests. According to Perkin–Elmer interviewees, scientist/users working at nonprofit institutions felt that free sharing of innovations was the established norm in the field.

Notes

1. Dr. H. H. Hausdorff, conversation with the author at Perkin–Elmer Corp., Norwalk, Conn., 1975.
2. M. Golay, in *Gas Chromatography 1957: Lansing Symposium,* ed. V. Coates, H. Noebels, and I. Fogerson (New York: Academic Press, 1958), 1–13.
3. M. Golay, in *Gas Chromatography 1958: Amsterdam Symposium,* ed. D. Desty (London: Butterworths, 1958), 36–55.
4. G. Dijkstra and J. de Gory, in D. Destry, ed., *Gas Chromatography 1958,* 36–55.

Silanization of Column Support Material

As mentioned earlier, gas chromatography is a physical means of separating chemical mixtures. Since any chemical reactions that occur within the instrument itself will cause false results, it is important that materials that come into contact with the sample be chemically inert with respect to it. A major step forward in rendering the granular column packing material (column support material) inert to important classes of sample material was a chemical treatment called silanization. Prior to the introduction of a quality-controlled silanized support material, investigators had great difficulty analyzing such materials as steroids, alkaloids, bile acids, and pesticides with the GC: These had tended to break down owing to chemical interaction with earlier support materials.

Development History

Silanization of column support material was developed by E. C. Horning, M. Horning, W. Van den Heuvel, and others, at the National Institutes of Health (NIH), Bethesda, Maryland, and at Baylor University College of Medicine, Houston, Texas. The work was performed in the 1958–64 period and resulted in numerous publications.

Commercialization

In the late 1950s and 1960s, E. C. Horning and his colleagues were involved with work on atherosclerosis at NIH. For their work they required special chemicals that were available from only two companies. One of these, Applied Science Libraries, State College, Pennsylvania, was a company founded by a former chemistry professor, Dr. Arthur Rose, of Pennsylvania State University. Horning became a customer of Applied Science Laboratories and a friendship developed between Horning and Rose. When Horning developed his novel column support material, he convinced the firm to enter the business of producing column supports.[1]

The problems involved in producing large batches of Horning's silanized column support material of consistent quality were addressed and solved by Applied Science Laboratories personnel. Gas-Chrome Q, the first commercially produced silanized column support material, was introduced to the market by Applied Science Laboratories in December 1964. The first competitors, F&M Scientific and Johns Manville, entered the market approximately one and a half years later.

Note

1. Dr. E. C. Horning, Baylor University College of Medicine, Houston, Tex., and R. Kruppa, Applied Science Laboratories, State College, Penn., telephone interviews, 1975.

Thermal Conductivity (TC) Detectors

When components of a chemical mixture being analyzed emerge from the partitioning column of a gas chromatograph, their presence must be detected. The first commercialized means of performing this task was the thermal conductivity detector. "Basically, a TC cell consists of a block (usually metallic) containing a cavity through which the carrier gas flows. A heated element (thermistor or resistance wire) is positioned in the cavity and loses heat to the block at a rate depending upon the TC of the gas. Since absolute measurement of TC is difficult, a differential means is usually employed. Thus, for example, two cavities can be drilled in the block, each one containing an element. Only carrier gas passes through one cavity and the column effluent passes through the other."[1] The difference in thermal conductivity is then measured in terms of the difference in electrical resistance of the two detectors.

Development History

The use of thermal conductivity for detection of gases is itself an old concept. It was apparently first applied to gas chromatography in 1954 by N. H. Ray, a scientist at ICI in England.[2] Ray visited the A.J.P. Martin and A. T. James laboratory (according to Martin, the originator of gas chromatography) and expressed an interest in examining compounds that were not detectable by titration. Martin suggested that Ray try a TC detector as this technique had already been used successfully by Caesson at Shell Amsterdam as a detector for his gas-solid chromatograms.[3]

Commercialization

Knowledge of thermal conductivity detection was widespread and detectors embodying this principle were installed on all of the early commercial gas chromatographs introduced around 1954 by firms such as Griffin and George, London, and Perkin–Elmer, Norwalk, Connecticut.

Notes

1. Alexander E. Lawson, Jr., and James M. Miller, "Thermal Conductivity Detectors in Gas Chromatography," *Journal of Gas Chromatography* 4 (August 1966): 273–84.
2. N. H. Ray, "Gas Chromatography: II. The Separation and Analysis of Gas Mixtures by Chromatographic Methods," *Journal of Applied Chemistry* 4 (February 1954): 82–85.
3. A. Martin, "The Past, Present and Future of Gas Chromatography," in *Gas Chromatography 1957: Lansing Symposium,* ed. V. Coates, H. Noebels, and I. Fogerson (New York: Academic Press, 1958), 237–47.

Argon Ionization Detector

The argon ionization detector is a great deal more sensitive to the presence of organic compounds than earlier detectors. Its operation is based on the observation by J. E. Lovelock that argon when used as a carrier gas could be ionized and then would transfer significant amounts of excitation energy to any organic vapors that were present. A beta ray ionization detector could then be used to sensitively detect any such transfer as a loss in ionization current. This in turn would signal the presence of the vapor.[1]

Development History

Lovelock, an English scientist without organizational affiliation, had been working with fatty acids in red blood cells and had difficulty in achieving good results with existing detectors used in gas chromatography. He turned his attention to this problem and developed the argon ionization detector around 1957.[2] Lovelock patented the innovation.

Commercialization

The argon ionization detector was first commercialized in 1959 by W. G. Pye and Company, an English manufacturer of gas chromatographs. Pye had begun production of GCs early on, 1954–55. Many of their customers, the petrochemical companies, had complained about problems with existing detectors (e.g., gas density balance and thermoconductivity). Pye knew of Lovelock's work with the new argon ionization detector, sought him out, and licensed his detector. Shortly after Pye's commercial introduction, other instrument makers began producing this detector by license from Lovelock.[2]

Notes

1. J. E. Lovelock, "A Sensitive Detector for Gas Chromatography," *Journal of Chromatography* 1 (1958): 35–46.

2. Dr. S. R. Lipsky, Yale Medical School, New Haven, Conn., telephone interview, 1975. (Lovelock and Lipsky collaborated in 1958–59 to develop the electron capture detector. Based on his acquaintanceship with Lovelock, Lipsky felt able to comment on the history of the argon ionization detector.)

Electron Capture Detector

The electron capture detector is an improved version of the argon ionization detector (see preceding entry). It has great sensitivity to materials such as pesticides, which are often present in samples in trace amounts only.

Development History

Lovelock and Lipsky met at a meeting of the New York Academy of Sciences in 1958–59 and decided to combine their efforts in order to improve GC detectors. Lovelock went to New Haven for six months as Lipsky was affiliated with the Yale Medical School. Together, they developed the electron capture detector.[1]

Commercialization

The electron capture detector was first commercialized by Barber-Coleman in 1960. Information on the device was transferred to Barber-Coleman by Lipsky who consulted for that firm. Shortly after the Barber-Coleman introduction, many firms began producing electron capture detectors.

Lipsky reports that his relationship with Barber-Coleman developed as a result of that firm being a supplier of GCs to his laboratory.[2]

Notes

1. J. E. Lovelock and S. R. Lipsky, "Electron Affinity Spectroscopy—A New Method for the Identification of Functional Groups in Chemical Compounds Separated by Gas Chromatography," *American Chemical Society Journal* 82, no. 1 (20 January 1960): 431–33.

2. Dr. S. R. Lipsky, Yale Medical School, New Haven, Conn., telephone interview, 1975.

Flame Ionization Detector

The flame ionization detector was introduced at approximately the same time as the argon ionization detector (see above) and provided similar major advantages over previously existing detectors used in gas chromatography. The detector measures the effect of organic vapors on the electrical conductivity of a flame.

Development History

The flame ionization detector was developed independently and nearly simultaneously by two groups: I. G. McWilliam and R. A. Dewar of ICI[1]; and J. Harley of the Engineering Research Section, South African Iron and Steel Industrial Corporation, with M. Nel and V. Pretorius of the University of Pretoria, South Africa.[2] There is some controversy over which group had priority. Patent rights, however, were obtained by ICI and licenses were issued by that firm to instrument manufacturers.

Commercialization

The flame ionization detector was described at an international symposium on gas chromatography in 1958[3] and also in an article in *Nature* in 1958.[4] Instrument firms immediately saw the commercial value of the idea and proceeded with commercialization even before the ICI patent was issued. The first firm to put the detector in the marketplace was apparently Perkin–Elmer Corporation in September 1959.

Notes

1. I. G. McWilliam and R. A. Dewar, "Flame Ionization Detector for Gas Chromatography," *Nature* 181, no. 4611 (15 March 1958): 760.
2. J. Harley, W. Nel, and V. Pretorius, "Flame Ionization Detector for Gas Chromatography," *Nature* 181, no. 4603 (18 January 1958): 177–78.
3. I. McWilliam and R. Dewar, in *Gas Chromatography 1958: Amsterdam Symposium,* ed. D. Desty (London: Butterworths, 1958), 142–45.
4. McWilliam and Dewar, "Flame Ionization Detector," 760.

Mass Spectrograph Detector

The mass spectrograph is an instrument that distinguishes molecules on the basis of their mass. It is an important instrument and widely utilized in its own right. Applying it as a detector to the gas chromatograph was a logical step but difficult to achieve—the gaseous output of a gas chromatograph is normally too large and too rapidly changing for a mass spectrograph to utilize. Several successful means (e.g., a device that separates the high volume of carrier gas from the sample gas) have been used to adapt the output of the gas chromatograph to the input requirements of the mass spectrograph.

Development History

The first successful linkage of a mass spectrograph and a gas chromatograph appears to be that of R. S. Gohlke, a researcher at Dow Chemical.[1] In 1957 Gohlke used a Bendix time of flight mass spectrograph owned by Dow as his detector. In 1960, L. P. Lindeman and J. L. Annis published a means to link the gas chromatograph to a magnetic mass spectrograph.[2] In the early 1960s Ryhage also developed and patented a jet separator linkage for this purpose.[3]

Commercialization

When Gohlke published his paper on the linkage he had accomplished between the gas chromatograph and the Bendix time of flight mass spectrograph, Bendix personnel heard about the achievement and produced a commercial version in January 1959. Their system was, however, expensive and difficult to operate and was not a commercial success. In the early 1960s Ryhage attempted to commercialize his linking device on his own and failed. A Swedish company, LKB, then obtained exclusive rights to the Ryhage patent and successfully commercialized the device in November 1965.

Notes

1. R. Gohlke, *American Chemical Society, Division of Petroleum Chemistry Preprints* 2, no. 4, D77–D83 (1957).
2. L. P. Lindeman and J. L. Annis, "Use of a Conventional Mass Spectrometer as a Detector for Gas Chromatography," *Analytical Chemistry* 32, no. 13 (December 1960): 1742–49.
3. Klaus Biemann, Professor of Chemistry, MIT, conversation at MIT, Cambridge, Mass., 1975. Professor Biemann developed a GC/MS linking device that has been

commercialized by both Consolodated Electrodynamics and Varian Associates, and he is intimately familiar with related work in this field.

Gas Sampling Valve with Loop

The gas sampling valve with loop is a simple innovation that had the effect of making the gas chromatograph into an instrument that could conveniently yield quantitative as well as qualitative results. To achieve this end it is necessary to inject a precisely known volume of the sample. The means chosen was the addition of a prechamber to the input valve of the gas chromatograph in the form of a loop of metal tubing. A gas sample of known volume and pressure was accumulated in this prechamber prior to injection into the gas chromatograph itself.

Development History

Investigators interested in achieving quantitative results with a gas chromatograph utilized various ad hoc means to this end. To my knowledge, however, no one made an addition to the instrument to accomplish that end conveniently and routinely prior to H. H. Hausdorff, E. Watson, and Bresky of Perkin–Elmer Corporation; they developed the gas sampling valve with loop at Perkin–Elmer[1] and assigned it to that firm. E. Watson received a patent for the innovation.[2]

Commercialization

Perkin–Elmer's gas chromatography group was motivated to develop the gas sampling valve with loop by the commercial possibilities associated with making the GC a more convenient instrument to use in quantitative studies. The same commercial assessment led them to quickly commercialize the valve, first making it available in 1956.

Notes

1. H. H. Hausdorff and L. J. Ettre, interview at Perkin–Elmer Corp., Norwalk, Conn., 1975.
2. U.S. Patent No. 2,757,541.

Process Control Chromatography

Gas chromatographs are extensively used in the on-line control of chemical process plants. Adopting the GC to this purpose required the development of special valves that would automatically sample a stream of gases moving through a plant. It also required the improvement of the long-term stability of GC performance: Users needed to be sure that a change in readings indicated a real change in plant performance rather than merely variations in performance of the monitoring instrumentation.

Developmental History

Early work in adapting GCs to process control was carried out by users in chemical and petroleum firms. An early U.S. application of a GC to this end was by researchers at Esso, who in 1956 reported on an application in an Esso refinery in Linden, New Jersey.[1] Another early developer was Keulemans, a researcher at Shell in the Netherlands.

Commercialization

Three companies were active in the introduction of process control gas chromatography: Beckman Instruments, Consolidated Electrodynamics, and Podbielniak Corporation. There is some uncertainty as to which of these firms was first on the market—probably Podbielniak was first. Seaton Preston, then at Podbielniak, reports that firm's first sale of a process control gas chromatograph to an Italian customer in 1956. The other two firms did not enter the market until 1957.[2] Podbielniak's equipment adopted the sampling valve developed by Keulemans of Shell. Initially, they built the valve in-house. Later, they turned to the Dutch firm that had fabricated valves for Keulemans.[3]

Notes

1. "Gas Chromatography Growing," *Chemical and Engineering News* 34, no. 15 (9 April 1956): 1692–96.
2. "Chromatography Reaches the Plant," *Chemical and Engineering News* 35, no. 33 (19 August 1957): 77–79.
3. Seaton Preston, telephone interviews, 1975. (Preston was an employee of Podbielniak Corp. at the time of the commercialization of process control gas chromatography.)

Preparative Gas Chromatography

Gas chromatographs used to identify chemicals in a mixture were only large enough to isolate a sufficient amount of each chemical to allow identification: They were not large enough to isolate sufficient material that would be useful in further chemical procedures. Since the material isolated by gas chromatography was very pure, it seemed desirable to scale up the process to the point at which useful amounts of chemical could be isolated through gas chromatographic techniques. Scaling up was achieved by increasing column size, by installing multiple columns, and by employing other means.

Development History

D.E.M. Evans and J. C. Tatlow at Birmingham University in England reported scaling up a GC column to separate fluorinated hydrocarbons in 1955.[1] Others soon reported other scale-up techniques to achieve preparative gas chromatography.[2]

Commercialization

The first firm to commercialize a preparative GC was Beckman Instruments Corporation, which introduced its Megachrom in March 1958. This device utilized larger-than-standard multicolumns that operated in parallel. A recycling system for the carrier gas (helium) was used for cost efficiency. D. Carle, one of the developers of the Megachrom at Beckman, advises that this scale-up approach was not copied from a user instrument but was original to Beckman.[3]

Notes

1. D.E.M. Evans and J. C. Tatlow, "The Reactions of Highly Fluorinated Organic Compounds. Part VIII. The Gas-chromatographic Separation on a Preparative Scale, and Some Reactions, of 3*H*- and 4*H*-Nonafluoro*cyclo*hexene," *Journal of the Chemical Society* (London) 1955: 1184–88.
2. D. Ambrose and R. R. Collerson, "Use of Gas-Liquid Partition Chromatography as a Preparative Method," *Nature* 177, no. 4498 (14 January 1956): 84.
3. D. Carle, Beckman Instruments Corp., telephone interview, 1975.

THE NUCLEAR MAGNETIC RESONANCE (NMR) SPECTROMETER

The nuclear magnetic resonance spectrometer identifies structural properties of molecules. It operates by simultaneously applying a steady magnetic field and a radio frequency signal to a sample of atoms or molecules. Adjustment of the frequency of the radiation and the strength of the magnetic field produces variations in the amount of radiation absorbed. The absorption is the result of NMR and occurs in spectra that are characteristic for different molecules and atoms and that can be used to identify them.

Development History

The original discovery of the NMR phenomenon was made by Felix Bloch, Professor of Physics at Stanford University, in 1946.[1] Bloch and E. M. Purcell, Professor of Physics at Harvard University, received a Nobel Prize for their research related to the phenomenon. The usefulness of the innovation to chemists was created with a discovery of the so-called chemical shift—the shift of nuclear magnetic resonant frequencies that occurs as a result of interactions with nearby atoms in a sample. This provides information about the chemical structure of organic molecules and other materials. The chemical shift phenomenon was discovered by W. G. Proctor and F. C. Yu in 1950.[2] The authors worked for Bloch at Stanford University. W. C. Dickinson[3] also played an important role in this discovery. The technique was further developed by H. S. Gutowsky.[4]

Commercialization

Prior to commercialization, there were only two NMR spectrometers in existence, those of Bloch at Stanford and Purcell at Harvard. Russell Varian, who had studied physics at Harvard, convinced Bloch to patent NMR, which Bloch did—receiving a very broad patent. Russell and his brother then took a license to the NMR patent, which they transferred to Varian Associates when they founded the company in April 1948. The first commercial NMR spectrometer was built by Varian Associates in 1950–51. The first three high-resolution NMR spectrometers commercialized were delivered to Humble Oil, Shell Development Corporation, and the Du Pont experimental station. These were operating by 1952.

The circuits used in Varian's first spectrometer had been published in the literature. As time went on, however, researchers employed by Varian Associates made significant contributions to the evolving field of NMR. Key figures in addition to the Varian brothers were James N. Shoolery, Emery Rogers, Forrest Nelson, Martin E. Packard, and Weston A. Anderson.[5]

In 1956 Perkin–Elmer introduced a low-resolution, broad-line spectrometer. This model, however, was quickly discontinued and Varian then held a virtual monopoly in high-resolution NMR until September 1966 when JEOL (a Japanese corporation) introduced their model C60H. In April 1967 Perkin–Elmer introduced their model R20, and in 1968 Brucker Magnetics entered the field by introducing their model HX60. Varian and Brucker Magnetics are now the industry leaders. JEOL is third.

Notes

1. F. Bloch, "Nuclear Induction," *Physical Review* 70, nos. 7–8 (1–15 October 1946): 460–74.
2. W. G. Proctor and F. C. Yu, "The Dependence of a Nuclear Magnetic Reso-

nance Frequency upon Chemical Compound," *Physical Review* 77, no. 5 (March 1950): 717.

3. W.C. Dickinson, "Dependence of the F^{19} Nuclear Resonance Position on Chemical Compound," *Physical Review* 77 (1 March 1950): 736–37.

4. H. S. Gutowsky and R. E. McClure, "Magnetic Shielding of the Proton Resonance in H_2, H_2O, and Mineral Oil," *Physical Review* 81, no. 2 (15 January 1951): 276–77.

5. F. Bloch, Professor of Physics, Stanford University, Dr. Emery Rogers, Hewlett-Packard Corp., and J. N. Shoolery and W. A. Anderson, both of Varian Assoc., telephone interviews, 1974.

Spinning of a Nuclear Magnetic Resonance Sample

Samples placed in a nuclear magnetic resonance spectrometer are subjected to a strong magnetic field. From a theoretical understanding of the NMR phenomenon, it was known by both NMR spectrometer users and personnel of the then-only manufacturer of NMR equipment (Varian Associates, Palo Alto, California) that increased homogeneity of the magnetic field would allow NMR equipment to produce more detailed spectra. Physical spinning of the sample within the spectrometer, the innovation being described here, is one way to improve the effective homogeneity of the field. Spin is achieved by linking a small electric motor or air turbine to an appropriately designed sample holder.

Development History

Felix Bloch (the discoverer of the NMR phenomenon) suggested rapid sample spinning as one means of improving effective magnetic field homogeneity and, thus, the resolution of NMR spectra.[1] Two of Bloch's students, W. A. Anderson and J. T. Arnold, built a prototype spinner and experimentally demonstrated the predicted result.[2]

Commercialization

Varian engineers went to Bloch's laboratory, examined his prototype sample spinner, developed a commercial model, and introduced it into the market by December 1954. The connection between Bloch and Varian was so good and Varian's commercialization of the improvement so rapid that there was little time for other users to construct homebuilt spinners prior to that commercialization.

Notes

1. F. Bloch, "Line-Narrowing by Macroscopic Motion," *Physical Review* 94, no. 2 (15 April 1954): 496–97.

2. W. A. Anderson and J. T. Arnold, "A Line-Narrowing Experiment," *Physical Review* 94, no. 2 (15 April 1954): 497–98.

Pulsed NMR Spectrometer

Applying magnetic pulses rather than a steady magnetic field to samples being analyzed in an NMR spectrometer allows one to examine molecular dynamics. Among the phenomena made accessible to observation by the technique are rates of reaction and molecular relaxation times.

Development History
The spin echo phenomenon underlying pulsed NMR was discovered—and patented—in 1950 by E. L. Hahn, Professor of Physics at the University of Illinois.[1] Numerous authors explored and refined its application to NMR over the succeeding years.[2] Some 12 users built their own equipment in the years before equipment embodying it was commercially available.

Commercialization
Interviewees at Varian say that the firm did produce "one or two" pulsed NMRs in 1954 but then did not pursue commercialization further. The Harvey Wells Corporation was the first to make a significant commercialization effort: They brought their ELH spin echo system to market in May 1964.

Professor John S. Waugh of MIT was reportedly responsible for kindling the interest of the Harvey Wells Corporation in pulsed NMR. He had come to know the company through purchasing their magnets to build his own pulsed NMRs. Waugh told the Harvey Wells Corporation of the concept and suggested they license Hahn's patent—which they did. Hahn and Waugh both consulted for Harvey Wells and aided their effort to develop a commercial device.[3] Pulsed NMR was next commercialized by NMR Specialties and Brucker Magnetics, both in 1962.

Notes

1. E. L. Hahn, "Spin Echoes," *Physical Review* 80, no. 4 (18 November 1950): 580–94.
2. See H. Y. Carr and E. M. Purcell, "Effects of Diffusion on Free Precession in Nuclear Magnetic Resonance Experiments," *Physical Review* 94, no. 3 (1 May 1954): 630–38; see also M. Sasson, A. Tzalmona, and A. Loewenstein, "A Spin Echo Attachment to Varian HR60 Nuclear Magnetic Resonance Spectrometer," *Journal of Scientific Instruments* 40 (1963): 133–34.
3. John S. Waugh, Professor of Chemistry, MIT, conversation at MIT, Cambridge, Mass., 1975.

Fourier Transform/Pulsed NMR

Fourier transform/pulsed NMR was an innovation applicable to the same analyses of molecular structure as conventional high-resolution NMR, but it offered an improvement in sensitivity of two orders of magnitude. The innovation was accomplished by coupling a computer capable of performing a fast Fourier transform algorithm directly to an NMR spectrometer.

Development History
Fourier transform/pulsed NMR was developed by R. R. Ernst and W. A. Anderson, two very creative researchers working at Varian Associates, a manufacturer of NMR spectrometers. They published their concept in 1966.[1] Development of a form of Fourier transform that was susceptible to rapid computation—a fast Fourier transform—by J. W. Cooley and J. W. Tukey in 1967[2] made the concept commercially feasible using the computers of the day. The product was developed in-house at Varian. Ernst and Anderson did not utilize the help of outsiders in their development work, but they were aware that other user and commercial groups were probably working on similar projects at the same time.

Commercialization

In July 1969 Varian reached the commercialization stage with the Ernst and Anderson project and introduced a Fourier transform accessory for its HA line of NMR spectrometers. This accessory was to be used in conjunction with an IBM 7090 computer. It was quickly followed by Digilab, which introduced an accessory in October 1969. Other quick commercializers were: Fabri-tek Instruments in February 1970, with a system based on a DEC PDP 8 computer; Brucker Magnetics in October 1970; and JEOL in 1971, with the PS–100 spectrometer with integrated computer.

Notes

1. R. R. Ernst and W. A. Anderson, "Application of Fourier Transform Spectroscopy to Magnetic Resonance," *Review of Scientific Instruments* 37, no. 1 (January 1966): 93–102.
2. William T. Cochran et al., G-AE Subcommittee on Measurement Concepts, "What Is the Fast Fourier Transform?" *IEEE Transactions on Audio and Electroacoustics* Au–15, no. 2 (June 1967): 45–55.

Heteronuclear Spin Decoupling

Heteronuclear spin decoupling is most useful for NMR spectroscopy work involving carbon 13 (C^{13}). Organic molecules have carbon as a "skeleton" with protons attached to them. This results in a very complicated NMR spectrum that can be simplified by heteronuclear spin decoupling.

Heteronuclear spin decoupling involves observing the C^{13} spectrum while irradiating the protons with a noise source tuned to their resonant frequency. This destroys the coupling of the protons to the C^{13} spectrum and thus makes that spectrum simpler. It also makes the C^{13} spectrum stronger (Overhauser effect), in effect increasing the sensitivity of the instrument in this application.

Development History

Double irradiation, the phenomenon at the base of heteronuclear spin decoupling, was first suggested by Felix Bloch in 1954. This basic idea was jointly developed and tested by Bloch with Martin E. Packard and James N. Shoolery at Varian Associates. The patent on the innovation is held by all three. In 1955 Arnold L. Bloom (also of Varian) and Shoolery published a simplification of the technique.[1]

Commercialization

Felix Bloch, Professor of Physics at Stanford University, and Varian Associates already had a close relationship established at the time of this innovation. In 1958 Varian commercialized a heteronuclear spin decoupler accessory. In 1960 Varian discontinued this accessory, and in 1962 they allowed a field engineer for Varian, a Mr. Yeiko, to offer the accessory in a new company he was founding, NMR Specialities. In 1962 Perkin–Elmer offered the capability. In 1964 Varian reintroduced the heteronuclear spin decoupling capability on its HA line of spectrometers. In 1966 JEOL (a Japanese corporation) offered it on its C60H model and in 1968 Brucker Magnetics offered it on their HF and HX series NMR spectrometers.

Note

1. Arnold L. Bloom and James N. Shoolery, "Effects of Perturbing Radiofrequency Fields on Nuclear Spin Coupling," *Physical Review* 97, no. 5 (1 March 1955): 1261–65.

Homonuclear Spin Decoupling

Homonuclear spin decoupling is used in proton spectroscopy. It allows one to associate particular lines in the spectrogram with particular protons. It is accomplished by irradiating one specific proton line while observing another proton line or group of lines. If irradiating the former affects the latter, then one knows that the two lines are closely related. If they were not, they would not display this coupling behavior.

Development History

Homonuclear spin decoupling goes back, as does heteronuclear spin decoupling, to a suggestion by Felix Bloch in 1954. W. A. Anderson of Varian Associates, however, was the first to demonstrate homonuclear decoupling of protons.[1] In 1959 Junkichi Itoh and Siro Sato[2] published a modification of the technique that considerably simplified the equipment required.

Commercialization

In 1964 Varian introduced homonuclear spin decoupling as an accessory on its HA60 and NMR spectrometer. James N. Shoolery of Varian estimates that between one- and two-dozen homemade versions of homonuclear spin decouplers were in use by users prior to commercialization. In 1967 Perkin–Elmer introduced homonuclear spin decoupling on its R20 NMR. In 1964 Brucker introduced it on its HX60 and JEOL introduced it on its C60 HL. In 1969 Varian introduced the innovation on its XL100 NMR spectrometer.

Notes

1. Weston A. Anderson, "Nuclear Magnetic Resonance Spectra of Some Hydrocarbons," *Physical Review* 102, no. 1 (1 April 1956): 151–67.
2. Junkichi Itoh and Siro Sato, "Double Proton Magnetic Resonance by a Side Band Method," *Journal of the Physical Society of Japan* 14 (1959): 851–52.

Frequency Synthesizer

The NMR spectrometer requires a very stable frequency source to operate properly. The most stable frequency sources commonly used in science incorporate quartz crystal references. Prior to the development of the frequency synthesizer, quartz crystal references were only available in the form of single-frequency oscillators. Each time one wanted to change frequencies, one had to change oscillators. The frequency synthesizer allows one to dial in on any frequency one wants with quartz stability—a major convenience for users.

Development History

Frequency synthesizers existed as commercial products prior to their application to NMR spectrometers by spectrometer manufacturers. Users who wished to have this feature on their NMRs simply purchased a synthesizer and wired it into their equipment appropriately.

Commercialization

In 1968 Brucker was the first to offer a frequency synthesizer built into their B–KR321S and B–KR322S models of NMR spectrometers. In September 1969 Varian offered a frequency synthesizer in their XL100 NMR spectrometers.

Superconducting Solenoids

The performance of NMR spectrometers improves as the magnetic field strength applied to the sample being analyzed is increased. Higher field strength increases both sensitivity and resolution of the instrument. Superconducting solenoids are capable of reaching significantly higher field strengths in the NMR application than are iron magnets. They are, however, somewhat less convenient to use as they must be cooled to liquid helium temperature for operation.

Development History

The NMR spectrometer with superconducting magnets was developed at Varian Associates, manufacturers of NMR spectrometers. Varian was induced to undertake the project by Bill Phillips of Du Pont. Phillips said that if Varian would build a superconducting NMR, Du Pont would place an order and provide some development money. Varian's development work involved research on field homogeneity, probe design for use with superconductors, and magnetic shielding as well as work on the superconducting magnets themselves.[1]

Commercialization

Varian introduced the NMR spectrometer with superconducting magnets in March of 1964. In February of 1965 Magnion offered a superconducting instrument appropriate for applications in which field homogeneity was not critical. In 1966 Varian introduced the HR–220 with a superconducting magnet system, and in 1971 Brucker introduced a superconducting magnet system.

Note

1. F. A. Nelson and H. E. Weaver, "Nuclear Magnetic Resonance Spectroscopy in Superconducting Magnetic Fields," *Science* 146 (9 October 1964): 223–32.

Shim Coils

To achieve high-quality NMR spectra it is important to subject the sample to a homogeneous magnetic field. Shim coils are a means of improving the homogeneity of the magnetic field. They consist of small electromagnetic coils attached to each pole face of the primary magnetics of the NMR spectrometer. The current flowing through these shim coils is carefully adjusted to correct any inhomogeneities detected in the magnetic field affecting the samples.

Development History

J. T. Arnold of Varian Associates was the first to propose the use of shim coils in NMR.[1] His paper appeared in 1956. In 1958 M. Golay, a scientist employed in Perkin–Elmer Corporation, another manufacturer of NMR spectrometers, published an improved shim coil design.[2] In 1959 H. Primas, Professor of Physics at the Swiss Federal Research Institute and a user of NMR, also published an article on shim coils.[3]

Commercialization

As far as we can determine, a Swiss firm, Trub–Tauber & Cie located in Zurich, was the first to commercialize NMR spectrometers with a shim coil, apparently in 1958.[4] Trub–Tauber & Cie obtained their design for shim coils from Professor Primas, an established consultant to that firm. Primas used NMR in his work and also was involved in designing NMR equipment. Varian Associates commercialized shim coils about 1960. In January 1966 Varian introduced a more sophisticated version of shim coils, Auto shim, on their HA100 model NMR spectrometer.

Notes

1. James T. Arnold, "Magnetic Resonances of Protons in Ethyl Alcohol," *Physical Review* 102, no. 1 (1 April 1956): 136–50.
2. Marcel J. E. Golay, "Field Homogenizing Coils for Nuclear Spin Resonance Instrumentation," *Review of Scientific Instruments* 29, no. 4 (April 1958): 313–15.
3. H. Primas, R. Arndt, and R. Ernst, *Zeitschrift für Instrumentenkunde* 67 (1959): 295.
4. Charles A. Reilly, "Nuclear Magnetic Resonance Spectrometry," *Review of Fundamental Developments in Analysis* 30, no. 4 (April 1958): 839–48.

Primas Polecaps

Primas polecaps are yet another means of improving homogeneity of magnetic field strength in the sample cavity of an NMR. The Primas design has the virtue of being of value for differing field strengths instead of being optimized for a particular field strength.

Development History

Primas polecaps are based on an idea by H. Primas, Professor of Physics at the Swiss Federal Research Institute. Primas was a user of NMR and also a consultant at Trub–Tauber & Cie, a manufacturer of NMR spectrometers.

Commercialization

In his role as consultant to Trub–Tauber & Cie, Primas transferred the design for Primas polecaps to that firm. When Brucker Magnetics, another manufacturer of NMR, acquired Trub–Tauber & Cie around 1964, they also acquired information regarding the Primas polecap design. Brucker commercialized these polecaps in 1968.

Field Frequency Lock

A field frequency lock keeps the ratio between the magnetic field and the radio frequency of an NMR spectrometer constant. The lock produces precisely correct spacing of absorption lines, which in turn allows exact superposition of sequential spectra for a given sample because, for each pass, the spectral lines are exactly aligned. This allows time-averaging experiments to improve signal-to-noise ratios when the signal is very weak. It therefore allows investigation of dilute samples and rare molecules.

Development History

After 1948 users of NMR knew that one could in principle build field frequency locks.[1] Since most users were chemists rather than electrical engineers, few actually did build such locks prior to the commercialization of the innovation. One of the users who did build such a lock was Saul Meiboom at Bell Laboratories.

Commercialization

In 1961 Varian Associates offered an external field frequency lock on their A60 and

NMR model. (An external field frequency lock is one in which the magnetic field sensing device is not quite in the same position as the sample. An internal field frequency lock is one in which the sensing device for the magnetic field is in the sample chamber itself.) In 1962 W. A. Anderson of Varian wrote an article on an internal field frequency lock control system,[2] and in February 1964 Varian was the first to commercialize an internal field frequency lock. In September 1966 JEOL (a Japanese corporation) offered both internal and external locks on their C60H NMR spectrometer. In July 1968 Brucker Magnetics commercialized internal and external field frequency locks on their HF60 model.

Notes

1. Martin E. Packard, "A Proton-Controlled Magnetic Field Regulator," *Review of Scientific Instruments* 19, no. 7 (July 1948): 435–39.
2. Weston A. Anderson, "Applications of Modulation Techniques to High Resolution Nuclear Magnetic Resonance Spectrometers," *Review of Scientific Instruments* 33, no. 11 (November 1962): 1160–66.

T_1*rho*

The T_1rho is used to measure slow motion in semisolids by measuring relaxation time. Earlier approaches to measuring slow motion involved lowering the magnetic field of the NMR instrument, but this caused signal-to-noise problems. The approach of the T_1rho accessory is to lower the amplitude of the radio frequency field. The radio frequency field then becomes the energy source that the material absorbs to relax. Maintaining high magnetic field strength during this process retains the high sensitivity of the NMR spectrometer.

Development History

T_1rho experiments were first done by physicists at the University of Illinois in a group working under Charles P. Slichter.[1] Numerous other groups did work on the problem over the next 10 years.[2] Although the original work was done on metal samples, interest grew among people doing research on biological materials and polymers.

Commercialization

In 1953 the T_1rho accessory was commercialized by Brucker Magnetics. As far as can be determined, the Brucker development group obtained the information they needed from published documents rather than through consultation with the developers of the technology.

Notes

1. Charles P. Slichter and David Ailion, "Low-Field Relaxation and the Study of Ultraslow Atomic Motions by Magnetic Resonance," *Physical Review* 135, no. 4A (17 August 1964): A1099–1110.
2. Brian D. Sykes and John M. Wright, "Measurement of Nuclear Spin Relaxation Times on an HA-100 Spectrometer," *Review of Scientific Instruments* 41, no. 6 (June 1970): 876–77. (These Harvard University authors describe modification of a Varian HA100 for the performance of T_1rho experiments.)

Pulsed Field Gradient Accessory

The pulsed field gradient accessory measures diffusion coefficients in semiliquid or liquid samples. It is an improvement on the pulsed NMR spin echo technique (see above). It involves reducing the field gradient during the times at which the radio frequency pulses are being applied to the sample and also at the time of appearance of the echo. This results in a broad echo that can be measured more accurately. The technique allows a one or two order-of-magnitude improvement in the accuracy of measurement of diffusion coefficients.

Development History

The pulsed field gradient accessory was developed by E. O. Stejskal and J. E. Tanner, researchers in the Department of Chemistry at the University of Wisconsin.[1] These authors had been using the spin echo method of E. L. Hahn[2] as developed by H. Y. Carr and E. M. Purcell.[3] Stejskal and Tanner were motivated to develop the pulsed field gradient accessory to eliminate some experimental limitations that they found in the spin echo method. Important precursors to their work were that of A. G. Anderson et al.[4] who noted diffusion in field gradients as well as that of D. W. McCall, D. C. Douglass, and E. W. Anderson[5] who noted the experimental possibilities of this technique in measuring self-diffusion coefficients. The basic apparatus used by Stejskal and Tanner[6] closely resembles that described by J. C. Buchta, H. S. Gutowsky, and D. E. Woessner earlier.[7]

Commercialization

The pulsed field gradient accessory was first commercialized by Brucker Magnetics. Brucker apparently acquired the information they needed to commercialize the unit from publications and conferences rather than consultants. Stejskal himself was not contacted by Brucker for help with the design.[8]

Notes

1. E. O. Stejskal and J. E. Tanner, "Spin Diffusion Measurements: Spin Echoes in the Presence of a Time-Dependent Field Gradient," *Journal of Chemical Physics* 42, no. 1 (1 January 1965): 288–92.
2. E. L. Hahn, "Spin Echoes," *Physical Review* 80, no. 4 (18 November 1950): 580–94.
3. H. Y. Carr and E. M. Purcell, "Effects of Diffusion on Free Precession in Nuclear Magnetic Resonance Experiments," *Physical Review* 94, no. 3 (1 May 1954): 630–38.
4. A. G. Anderson et al., "Spin Echo Serial Storage Memory," *Journal of Applied Physics* 26, no. 1 (November 1955): 1324–38.
5. David W. McCall, Dean C. Douglass, and Ernest W. Anderson, "Self-Diffusion Studies by Means of Nuclear Magnetic Resonance Spin-Echo Techniques," *Deutsche Berichte der Bunsen-Gesellschaft für Physikalische Chemie* 67, no. 3 (1963): 336–40.
6. E. O. Stejskal, "Use of an Analog-to-Digital Converter in Pulsed Nuclear Resonance," *Review of Scientific Instruments* 34, no. 9 (September 1963): 971–75.
7. J. C. Buchta, H. S. Gutowsky, and D. E. Woessner, "Nuclear Resonance Pulse Apparatus," *Review of Scientific Instruments* 29, no. 1 (January 1958): 55–60.
8. E. O. Stejskal, Department of Chemistry, University of Wisconsin, telephone interview, 1975.

Electronic Integrator

The electronic integrator integrates the area under the peaks of the NMR spectrometer spectrum. This area is proportional to the total number of protons contributing to that peak. It allows a much more accurate count of this number than was possible previously.

Development History

Integrators are used for a range of purposes in scientific research and were available as modules in the commercial marketplace prior to their commercialization as a built-in feature of an NMR spectrometer. Interested users could purchase these units and wire them into their equipment if they so desired. Among the users who did this were people such as Baker at Dow Chemical Research.[1]

Commercialization

J. N. Shoolery and W. A. Anderson were the researchers at Varian Associates who developed that firm's electronic integrator. Varian offered it as an accessory in 1960. In 1961 the Varian model A60 and NMR spectrometer had a built-in integrator. In 1965 JEOL offered an electronic integrator.[2]

Notes

1. J. N. Shoolery, Varian Associates, Palo Alto, Calif., telephone interview, 1975.
2. W. A. Anderson, Varian Associates, Palo Alto, Calif., telephone interview, 1975.

Multinuclei Probe

The multinuclei probe allows a user to examine any of more than 70 nuclei types without changing any hardware. The major advantage this innovation presents to users relative to old-style probes can be appreciated when one understands that such probes were restricted to the examination of only a single nucleus type and cost several thousand dollars each. Also, the time involved in changing from one probe to another could take several hours.

Development History

The multinuclei probe was developed by Daniel Traficante, a visiting scientist in MIT's Department of Chemistry, and by Michael Mulcay, president of United Development Corporation in Lexington, Massachusetts. The two lived in the same town and knew each other socially. One day Traficante told Mulcay, an electronics engineer, about his idea for a multinuclei probe, and they decided to get together to work on it. The first probe they developed was installed at MIT in 1972. The second was installed at Michigan State University in 1974; one was also installed in Australia in 1974. All of these multinuclei probes were built and installed by Traficante and Mulcay according to a design they published in September 1974.[1]

Commercialization

Interest in multinuclei probes on the part of users apparently stimulated Varian Associates to develop their own multinuclei probe in cooperation with Paul Ellis of the University of South Carolina. This probe design was commercialized in 1974. In late

1974 Brucker Magnetics commercialized a multinuclei probe of the Traficante and Mulcay design. The arrangement at that time was that Brucker would supply Traficante and Mulcay with parts and components and that those two would build the probes and would sell them back to Brucker for marketing.

Note

1. Daniel D. Traficante, James A. Simms, and Michael Mulcay, "An Approach to Multinuclei Capability in Modern NMR Spectrometers," *Journal of Magnetic Resonance* 15 (September 1974): 484–97.

Proton-Enhanced Nuclear Induction Spectroscopy

Proton-enhanced nuclear induction spectroscopy allows the examination of the molecular structure of molecules that either are naturally rare or that have extremely small magnetic moments. When a nucleus is rare, the signal is weak. The normal solution to weak signals (time averaging) cannot be used here because the relaxation times of the spins are too long. That is, the spin takes too long to recover its original polarization, which is lost during an observation—the relaxation time may be one hour. The proton-enhanced nuclear induction spectroscopy method solves this problem by replenishing the polarization of the rare spins from the abundant spins rather than in the usual way through the lattice. This makes use of the coupling between rare and abundant spins in the way proposed by S. R. Hartmann and E. L. Hahn.[1] One establishes the coupling by double resonance and polarizes the rare spins in (perhaps) a millisecond. In effect, one pumps the polarization of the abundant spins out through the rare spins and thus elicits their spectra. The abundant spins are usually protons.[2]

Development History

S. R. Hartmann and E. L. Hahn, Department of Physics, University of California, Berkeley, introduced the double resonance method in 1962. F. M. Lurie and C. P. Slichter, University of Illinois, performed a double resonance experiment on lithium metal and carried it further in 1964.[3] In 1965 R. Freeman and W. A. Anderson published their experiments with double resonance that used modulated perturbing radio frequency fields.[4]

Commercialization

Brucker Magnetics was induced to work on commercializing proton-enhanced nuclear induction spectroscopy by John S. Waugh, Professor of Chemistry, MIT. Waugh was a consultant to Brucker. Varian Associates also initiated work on the technique by 1973.

Notes

1. S. R. Hartmann and E. L. Hahn, "Nuclear Double Resonance in the Rotating Frame," *Physical Review* 128, no. 5 (1 December 1962): 2042–53.

2. John S. Waugh, Professor of Chemistry, MIT, conversation at MIT, Cambridge, Mass., 1974.

3. Fred M. Lurie and Charles P. Slichter, "Spin Temperature in Nuclear Double Resonance," *Physical Review* 133, no. 4A (17 February 1964): A1108–22.

4. R. Freeman and W. A. Anderson, "Nuclear Magnetic Double Resonance. Transmission of Modulation Information Through the Nuclear Spin-Spin Coupling," *Journal of Chemical Physics* 42, no. 4 (15 February 1965): 1199–1229.

THE TRANSMISSION ELECTRON MICROSCOPE (TEM)

The First Electron Microscope

The electron microscope is the electronic analog to the optical microscope. The conventional optical microscope uses glass lenses to bend and focus beams of light to create an enlarged image of a sample. Analogously, the electron microscope uses magnetic and electric fields to bend a beam of electrons to achieve an enlarged image of a sample. Because the wavelength of electrons is more than 1000 times shorter than the wavelength of light used in conventional light microscopy, the electron microscope can achieve 1000 times greater resolution than can the light microscope. Although the optical microscope is limited in resolution to about 1000 angstroms (Å), the resolution of a modern electron microscope can be on the order of 1Å.

The very great improvement in resolving power offered by the electron microscope over the best previously available optical microscopes made it enormously important to researchers in many fields, ranging from metallurgy to biology.

Development History

The first work in the electron microscope was done at the Technische Hochschule in Berlin by Max Knoll and his student Ernst Ruska. These two were working on high-speed oscilloscopes for the purpose of building devices capable of measuring the effects of lightning more effectively.[1] Their investigations of the optics of electrons in a vacuum led to their development of the electromagnetic lens, a key element in the electron microscope. In 1931 they reported their observation of the enlargement of the images of apertures.[2]

A few days after Knoll reported his initial results, G. R. Rudenberg, an employee of Siemens, a German electrical manufacturer, applied for a patent on the theory of the electron microscope.[3] The close timing of these two events plus the apparent lack of prior work on the part of Rudenberg that could be logically linked to an electron microscope caused some skepticism regarding the source of his insight. Rudenberg's patent covered the basic use of magnetic and electrostatic lenses to form electrooptical instruments in which electrons took the same part as light does in light optics. Also in 1931 E. Bruche (of the German firm AEG) published a report of work in electron optics.[4] By 1932 Knoll and Ruska had built an improved electron microscope that operated with an electron beam of 65 kv and reached 120 times magnification.[5]

L. Marton, a researcher in Belgium, had read some of Knoll and Ruska's early papers and started to build his own electron microscope in 1932. Marton focused on the problem of preparing samples to be observed under the electron microscope, especially biological samples. It had early been thought that an electron beam would destroy biological samples, but Marton demonstrated that, indeed, they could be prepared and examined.[6]

In 1933 Knoll and Ruska had improved their electron microscope to the point that it was able to exceed the resolution of an optical microscope. Biologists and chemists at the Technische Hochschule in Berlin made use of this microscope and became very interested in its possibilities. In 1935 Marton built an electron microscope that was able to exceed the resolution of an optical microscope by a factor of 10.

Commercialization

In 1937 Ruska and B. von Borries were hired by Siemens to develop an electron microscope for production. They had a prototype ready by the end of 1938, and in 1939

Siemens started production of commercial electron microscopes. In total, Siemens produced about 30 units during 1939 and the early war years.[7] In 1938 Marton left Europe to come to RCA in the United States, and he brought along with him an electron microscope of his own design.[8] In 1940 RCA also hired J. Hillier, another early researcher in electron microscopy. Before the end of 1941 Marton left RCA for Stanford University. In 1941 RCA commercialized a microscope of Hillier's design, which was known as the RCA Type B or the EMB.

Notes

1. Martin M. Freundlich, "Origin of the Electron Microscope," *Science* 142, no. 3589 (11 October 1963): 185–88.
2. L. Marton, *Early History of the Electron Microscope* (San Francisco: San Francisco Press, 1968), 5.
3. T. Mulvey, "Origins and Historical Development of the Electron Microscope," *British Journal of Applied Physics* 13 (1962): 197–207.
4. Marton, *Early History of the Electron Microscope,* 19.
5. Mulvey, "Origins and Historical Development of the Electron Microscope," 200–1.
6. Marton, *Early History of the Electron Microscope,* 19–22; and Mulvey, "Origins and Historical Development of the Electron Microscope," 202–4.
7. Mulvey, "Origins and Historical Development of the Electron Microscope," 206; and Ralph W. G. Wyckoff, *Electron Microscopy: Technique and Applications* (New York: Interscience, 1949), 14.
8. Marton, *Early History of the Electron Microscope,* 35.

Pointed Filaments

The electron microscope operates by accelerating a beam of electrons. The coherence of this beam of electrons is very important to achieving high resolution in the electron microscope, especially in the region below 8 Å resolution. The source of electrons in an electron microscope is a filament much like the filament in an electric light bulb. It was found that making this filament in a pointed shape greatly improved the coherence of the electron microscope beam.

Development History

Pointed filaments were developed in the early 1960s by H. Fernandez-Moran, a biologist at the University of Chicago.[1]

Commercialization

Various versions of pointed filaments were commercialized in the late 1960s by two firms that specialized in electron microscope accessories and supplies. These two firms were the C. W. French Company and the Ladd Company. We have little information on how these firms got involved in the production of pointed filaments. Ladd reported[2] that he visited the Fernandez-Moran laboratory and observed the fabrication of pointed filaments there.

Notes

1. T. Hibi, "Operating Condition of Point Cathode and Resolution of Electronmicroscope," in *Electron Microscopy 1964* (Proceedings of the Third European Regional Conference, Prague, 26 August–3 September 1964), ed. M. Titlbach, Vol. A

(Prague: Publishing House of the Czechoslovak Academy of Sciences, 1965), 121–22; H. Fernandez-Moran, "Applications of Improved Point Cathode Sources to High Resolution Electron Microscopy," in *Electron Microscopy 1966* (Sixth International Congress for Electron Microscopy, Kyoto, Japan, 28 August–4 September 1966), Vol. 1: *Non-Biology,* ed. Ryozi Uyeda (Tokyo: Maruzen, 1966), 27–28.

2. Mrs. Margaret Ladd, Burlington, Vt., telephone interview, 1974; C. W. French, Agawam, Mass., telephone interview, 1974.

Telefocus Electron Gun

The telefocus gun was a magnetic and electronic innovation that allowed the user of an electron microscope to separately control the diameter and the intensity of the electron beam of an electron microscope. Separate control of these functions is important to users, but they had been inherently linked in earlier gun designs.

Development History

The telefocus gun was designed by K. H. Steigerwald, a researcher at the Max Planck Institute in Germany.[1] Steigerwald reported his development in 1949. His design was quickly replicated by several metallurgists who had great need for it in their studies that required the use of the electron microscope.

Commercialization

Steigerwald was a consultant for Siemens, a German producer of electron microscopes. Siemens introduced the telefocus design in 1951. In the late 1950s Philips Corporation, Eindhoven, the Netherlands, commercialized it, and in 1965 RCA started selling a kit to retrofit their existing microscopes with the innovation.

Note

1. K. H. Steigerwald, "Ein Neuartiges Strahlerzeugungs-System für Elektronenmikroskope," *Optik* 5, no. 8/9 (1949): 469–78.

Double Condenser Lens

The double condenser lens replaces a single condenser lens in the optical system of an electron microscope. This substitution provides the user with several important improvements. First, the intensity of the electron beam can be varied over a wider range. This permits a greater range of specimen-viewing conditions for users: Strong illumination is possible for thick specimens, weak illumination is possible for delicate specimens. Second, the double condenser lens creates more working space between the condenser and objective lenses. This is the space in which users place their sample. The increase in space allows users to design accessories for handling samples with greater freedom. Finally, the double condenser lens enables users to reduce the area of their sample that is exposed to the electron beam. This in turn allows users to reduce specimen damage of delicate specimens.

Development History

L. Marton claims that he developed the double condenser lens at Stanford University in the 1940s.[1] It is possible that this was also done independently by other users since double condenser lenses were also a feature of optical microscopes and the analogy was clear.

Commercialization
The double condenser lens was first offered commercially by Siemens about 1952.[2] Several other microscope makers quickly followed, but RCA delayed until 1959, when they introduced a double condenser retrofit kit for their EMU3 electron microscope.[3]

Notes

1. L. Marton, *Early History of the Electron Microscope* (San Francisco: San Francisco Press, 1968), 41; also L. Marton, "A 100-kv Electron Microscope," *Journal of Applied Physics* 16 (March 1945): 131–38.
2. B. v. Borries, "The Physical Situation and the Performance of High-Resolving Microscopy Using Fast Corpuscles," in *Proceedings of the Third International Conference on Electron Microscopy, London 1954* (London: Royal Microscopical Society, 1956), 4–25, esp. 4 and Plate V.
3. John Coleman, former employee of RCA's Electron Microscope Division, conversations at MIT, Cambridge, Mass., 1974.

Correction of Astigmatism in the Objective Lens

Electron microscope lenses are subject to imperfections (e.g., astigmatism) just as optical lenses are. Astigmatism in the objective lens is the result of imperfections in the pole piece of the electromagnet used in that lens.

Development History
Many users were trying to improve the resolution of their microscopes by (among other things) polishing, boring, and mechanically adjusting the pole pieces of their objective lenses in order to eliminate astigmatism.[1]

Commercialization
The first to commercialize an adjustable astigmatism corrector for an objective lens was RCA on their advanced EMU model in 1947. The correction was achieved by eight screws built into the objective pole piece. This commercial version was developed by J. Hillier and E. G. Ramberg at RCA.[2]

Notes

1. Dr. John Riesner, RCA Sarnoff Laboratories, conversations at Cherry Hill, N.J., 1974.
2. James Hillier, "Further Improvement in the Resolving Power of the Electron Microscope," *Journal of Applied Physics* 17 (April 1946): 307–9; also James Hillier and E. G. Ramberg, "The Magnetic Electron Microscope Objective: Contour Phenomena and the Attainment of High Resolving Power," *Journal of Applied Physics* 18 (January 1947): 61–71.

Well-Regulated High-Voltage Power Supplies

Electron microscopes require high voltages to operate. A very high stability in the high-voltage supply of an electron microscope is a well-known prerequisite for achieving high resolution.

Development History
The first electron microscope and the first few precommercial replications used batter-

ies connected in series to supply the high voltages required. A very stable high voltage could be obtained by this means, but the major inconvenience associated with this solution can be readily imagined: voltages on the order of 80,000 v were needed—supplied by nearly 40,000 single wet-cell batteries connected in series. A visitor to the laboratory of L. Marton, an early and outstanding experimenter in electron microscopy, recalls an entire room filled with batteries on floor-to-ceiling racks with a full-time technician employed to maintain them. The first commercial electron microscope, built by Siemens in Germany in 1939, substituted a power supply for the batteries but could not make its output voltage as constant as could be done with the batteries.[1]

Commercialization

When RCA decided to build an electron microscope, an RCA electrical engineer, Jack Vance, undertook to build a highly stable power supply and by several inventive means achieved a stability almost good enough to eliminate voltage stability as a constraint on the performance of a high-resolution microscope. Vance achieved this by using radio frequency rather than line-voltage frequency to actuate the high-voltage rectifier tubes in the power supply and by incorporating a reference battery in a feedback circuit to stabilize the high voltage.[2]

Vance's innovative power supply was commercialized in 1941 in RCA's first-produced electron microscope, the RCA Type B (EMB).

Notes

1. Ralph W. G. Wyckoff, *Electron Microscopy: Technique and Applications* (New York: Interscience, 1949), 16.
2. A. W. Vance, "Stable Power Supplies for Electron Microscopes," *RCA Review* 5, no. 3 (January 1941): 293–300.

Regulated Lens Power Supply

The lenses of electron microscopes are electromagnets. To provide a steady magnetic field they require a very stable supply of electric current. This current is provided by the lens power supply. Early electron microscopes built by users and the early Siemens's commercially produced electron microscope used batteries in conjunction with water cooling as their lens power supply system. The batteries supplied the current necessary, while the water cooling prevented the resistance in the lens wiring from changing with temperature, thus changing current flow in the lens magnetic field. As can be imagined, both the batteries and the water cooling contributed significantly to the cumbersomeness and expense of the early electron microscopes.

Development History

Jack Vance of RCA eliminated both the need for batteries and the need for water cooling in the lens power supply he designed. He replaced the high-current, low-voltage battery supply with a low-current, high-voltage supply from a regulated power supply and designed a current regulator into this power supply, so that changes in temperature of the lens windings would not affect lens performance.[1]

Commercialization

The regulated lens power supplies designed by Vance were commercialized in the RCA Type B (EMB) electron microscope introduced commercially in 1941.

Note

1. A. W. Vance, "Stable Power Supplies for Electron Microscopes," *RCA Review* 5, no. 3 (January 1941): 293–300.

Rubber Gasket Sealing of Vacuum System

The interior of an electron microscope is kept under a very hard vacuum. This is essential to its operation as electron beams cannot travel effectively in air. The electron microscope has many joints that are opened and closed frequently by the operator. In modern electron microscopes, rubber gaskets are used to make these joints airtight when they are closed.

Development History

Early electron microscopes used carefully machined surfaces covered with a special grease to achieve a vacuum-tight seal. Although this procedure had often been followed in high-vacuum work of the period, it proved very troublesome in the case of electron microscopy. Sealing grease from the joints infiltrated the vacuum chamber of the electron microscope, where it became charged by the electron beam. This charged material then distorted the electron beam in various unanticipated ways, destroying the microscope's performance. L. Marton realized the problem and designed rubber gaskets into the electron microscopes he built in the late 1930s.[1] However, Siemens's 1939 production unit—the first electron microscope commercialized—used machined surfaces and vacuum grease.

Commercialization

When Marton came to RCA in 1938, he carried with him one of his microscopes that used rubber gaskets to seal the vacuum system.[2] This feature was adopted for the RCA Type B (EMB) electron microscope commercialized in 1941.

Notes

1. L. Marton, *Early History of the Electron Microscope* (San Francisco: San Francisco Press, 1968), 22.
2. Ibid., 35.

Three-Stage Magnification

Creating a three-stage magnification system for the electron microscope involved inserting a third intermediate lens in the electron optical system. Three-stage magnification allowed users to adjust the magnification more widely. It thus allowed them to examine the total specimen at one time at a low magnification.

Development History

In 1944 an electron microscope containing three-stage magnification was built by J. B. le Poole, Director of the Institute for Electron Microscopy in Delft, the Netherlands. The microscope was used for a month and then taken apart and hidden from the German enemy until the end of the war. After the war it was described in a *Philips Technical Review*.[2] An electron microscope with three-stage magnification was also built by L. Marton at Stanford University in 1944.[3]

Commercialization
J. B. le Poole, designer of the electron microscope containing three-stage magnification, was a consultant to Philips Corporation in Eindhoven, the Netherlands. In 1947 Philips introduced commercially a 100 kv electron microscope patterned after the microscope built at Delft.

Notes

1. L. Marton, *Early History of the Electron Microscope* (San Francisco: San Francisco Press, 1968), 41.
2. J. B. le Poole, "A New Electron Microscope with Continuously Varying Magnification," *Philips Technical Review* 9, no. 2 (1947): 33–45.
3. "A New Electron Microscope for 100 kV," *Philips Technical Review* 9, no. 6 (1947): 179.

Scaled-up Objective Pole Piece

A pole piece is part of an electromagnetic lens like those used in electron microscopes. An objective pole piece is, therefore, part of the objective lens of the electron microscope. The innovation of enlarging the objective pole piece had several advantages over previous pole piece design. It reduced spherical aberration—a type of flaw in lens performance—and it increased intensity of the image. It also reduced the cost of producing the lens by reducing both the machining accuracy needed and the precision with which the pole piece must be aligned in the electron microscope.

Development History
The scaled-up objective pole piece was developed by L. Marton when he was at RCA in 1940.[1] He incorporated it in his model No. 4, which was also known as the RCA Type A. Marton, in his 1968 book on the early history of the electron microscope, reports that this innovation was partly inspired by two theoretical papers, one by R. Rebsch and W. Schneider and the other by W. Glaser.[2]

Commercialization
Although the scaled-up objective pole piece was developed at RCA, it was first commercialized by Metropolitan Vickers Electrical Company in 1949 in their model EM-3[3]; Philips followed in 1950.[4]

Notes

1. L. Marton, M. C. Banca, and J. F. Bender, "A New Electron Microscope," *RCA Review* 5, no. 2 (October 1940): 232–43.
2. L. Marton, *Early History of the Electron Microscope* (San Francisco: San Francisco Press, 1968), 38.
3. M. E. Haine, R. S. Page, and R. G. Garfitt, "A Three-Stage Electron Microscope with Stereographic Dark Field, and Electron Diffraction Capabilities," *Journal of Applied Physics* 21 (February 1950): 173–82.
4. A. C. van Dorsten, H. Nieuwdorp, and A. Verhoeff, "The Philips 100 kV Electron Microscope," *Philips Technical Review* 12, no. 2 (August 1950): 33–64.

Goniometer Specimen Stage

A specimen stage is a device used to mount specimens properly within the electron microscope. A goniometer stage is a device that can be tilted in one or two directions while it is in the microscope. This feature is useful in many instances and is necessary in certain materials science work.

Development History

The earliest microscope stages were simple screens upon which the sample to be observed was placed. As electron microscopes and microscopy matured starting in the early 1960s, users started to need stages that would allow them to observe their samples under various special conditions. U. Valdre was apparently either the first or among the first to develop goniometer type stages. He was a materials scientist at the University of Bologna and needed such a stage for his work.[1]

Commercialization

Siemens and Philips commercialized Valdre-type goniometer stages in the early 1960s. RCA commercialized such stages in 1964.

Note

1. U. Valdre, "A Universal Specimen Stage and Combined Cartridges for an Electron Microscope," in *Electron Microscopy 1966* (Sixth International Congress for Electron Microscopy, Kyoto, Japan, 28 August–4 September 1966), Vol. 1: *Non-Biology*, ed. Ryozi Uyeda (Tokyo: Maruzen, 1966), 165–66.

Cold Specimen Stage

A cold specimen stage is one that can keep samples cold while they are being examined under the electron microscope. Cooling can be achieved quite simply by thermally connecting the stage to a source of liquid nitrogen outside the electron microscope.

Development History

Cold stages were developed very early in the history of electron microscopes by L. Marton.[1] Both the microscope Marton brought to RCA when he became an employee and the one he developed at RCA had such stages.

Commercialization

In the mid-1950s Siemens was first to commercialize the cold stage.[2] In the mid-1960s demand for such stages began to pick up among biologists who were eager to reduce the radiation damage inflicted on their samples by electron microscopy.

Notes

1. L. Marton, "A New Electron Microscope," *Physical Review* 58 (1 July 1940): 57–60.

2. F. S. Sjöstrand and J. Rhodin, eds., *Electron Microscopy, Proceedings of the Stockholm Conference September 1956* (New York: Academic Press, 1957), 27.

High-Temperature Specimen Stage

A high-temperature stage can heat up to perhaps 3000°C—a very high temperature—while inside the microscope. It is important that the electron microscope be designed in such a way that the expansion of metals, and so on, associated with the high temperatures not deform the microscope or degrade its resolution.

Development History

The first high-temperature specimen stage was developed by M. von Ardenne, a researcher at an institute in Berlin.[1]

Commercialization

In the late 1950s metallurgists started to demand high-temperature stages from manufacturers. In 1965 (approximately) RCA commercialized such a stage.

Note

1. Abstract in V. E. Cosslett, ed., *Bibliography of Electron Microscopy* (London: Edward Arnold, 1950), 18.

Biased Electron Gun

The device in the electron microscope that generates the electron beam used for "seeing" the sample is called an electron gun. It is important to the performance and resolution capability of an electron microscope that the electron gun generate an intense, stable, and coherent electron beam. On all these dimensions, the biased electron gun represented an improvement over the basic thermionic emission gun used in the first electron microscopes commercialized by Siemens and RCA.

Development History

The biased electron gun was developed by A. Wehnelt.

Commercialization

The biased electon gun was first commercialized by Siemens.[1]

Note

1. Ralph W. G. Wyckoff, *Electron Microscopy: Techniques and Applications* (New York: Interscience, 1949), 16.

Out-of-Gap Objective Lens

The specimen to be examined by electron microscope must be mounted in, or near, the so-called objective lens of the microscope. An effective design for such a lens will not only have superior electrooptical qualities, it will also have a geometry that allows users to surround the specimen with any needed experimental equipment. The out-of-gap lens performed well on both dimensions. Its unique feature was a geometry that allowed one to place the specimen entirely to one side of the confines of the lens, that is, out-of-gap.

Development History

In 1935 E. F. Burton, who had seen the electron microscope work being done in

Berlin, came to the University of Toronto, Canada, and transferred the idea of an electron microscope to that university. J. Hillier and A. Prebus, working for Burton in Toronto, built successful electron microscopes.[1] Hillier brought the knowledge of that work with him when he became an employee of RCA in 1940. At RCA Hillier built an electron microscope modeled on the design developed by the Toronto group. This design included an out-of-gap objective lens like that used in Toronto.

Commercialization

The out-of-gap objective lens developed in Toronto was commercialized by RCA in their first electron microscope, the RCA Type B of 1941. RCA continued to use this lens design in its commercial microscopes until about 1968.

Note

1. E. F. Burton, J. Hillier, and A. Prebus, "A Report on the Development of the Electron Supermicroscope at Toronto," *Physical Review* 56 (1 December 1939): 1171–72; see also Albert Prebus and James Hillier, "The Construction of a Magnetic Electron Microscope of High Resolving Power," *Canadian Journal of Research* 17A, no. 4 (April 1939): 49–63.

DATA SET FOR SEMICONDUCTOR PROCESS INNOVATIONS

The innovations described in this data set are innovations in the process by which silicon-based semiconductors are manufactured. Most of the innovations are embodied in novel process machinery, but a few are embodied in novel techniques carried out largely by hand.

Selection criteria for this sample will be found in chapter 2. Owing to these criteria, a few important innovations familiar to many readers (e.g., float zone refining) are not represented. In addition, we were unable to find any information on the innovation histories of four innovations included in our sample. These were coded NA in all relevant chapter 2 tables.

Growth of Single Silicon Crystals: The Crystal Puller

Silicon semiconductors—whether individual transistors or integrated circuits—are built up on a wafer sliced from a single crystal of silicon. It is crucial to the performance of the semiconductor that the single silicon crystal used be quite regular in structure and quite free from impurities. The method used for making pure single crystals of silicon for commercial semiconductor manufacture is called crystal pulling. It involves bringing a small seed crystal of silicon into contact with the surface of a bath of molten silicon. Molten silicon in contact with the small crystal is cooled enough to crystallize and thus extend that crystal. Gradual pulling of the seed crystal away from the molten material as crystallization occurs results in a long single crystal rod of pure silicon being grown over a period of time.

Development History

The method of growing single crystals by pulling from a melt dates back to the work of J. Czochralski in 1917.[1] Indeed, the method is sometimes called the Czochralski method. The method was later used by numerous researchers studying single crystals and became well-known in the field.

When Gordon Teal of Bell Laboratories began to study the fabrication of single crystals for use in semiconductors, he decided to adopt the crystal-pulling approach. He and his colleagues first made single crystal germanium by this technique and then began to work on the pulling of silicon single crystals. This proved a very difficult technical task since silicon melts at 1420°C, a temperature at which the containers holding the molten silicon tend to contribute impurities to the melt.

In January 1953 Teal took a job with Texas Instruments, where he formed a materials laboratory. He continued his work on the pulling of single silicon crystals with the aim of producing a commercially usable process. He succeeded, and in June 1954 Texas Instruments used the method in the production of the first commercial silicon transistor.

Commercialization

In the last two weeks of April 1952, Bell Laboratories held a symposium for 34 or so companies who had paid $25,000 each to become licensees of all Bell's proprietary semiconductor knowledge. Central to this information was the technique of crystal pulling. Since the symposium included laboratory visits, demonstrations of equipment, and so on, engineers of licensee companies learned enough to go and build their own crystal pullers. Bell also privately printed a book, *Transistor Technology,* which was distributed to licensees in September 1952.[2] After Bell's 1952 licensing conference, firms interested in manufacturing semiconductors began to build the crystal pullers they needed in-house. Participants in the industry at that time suspected that user firms may have called on outside manufacturers for help in fabricating the equipment to their designs, but they cannot recall the names of any such firms. The first firm we can identify that manufactured the Bell-designed crystal puller as a commercial product was a small firm called Lepel Corporation in Maspeth, New York. According to interviewees at that firm, they built a prototype puller for Bell Laboratories in 1956.

Notes

1. J. Czochralski, "A New Method for the Measurement of the Velocity of Crystallization of the Metals," *Zeitschrift für Physikalische Chemie* 92, no. 2 (1917): 219–21.
2. *Transistor Technology* was privately printed by Bell Laboratories in September 1952. An updated version (with the same title) was published by Van Nostrand, New York, 1957.
3. J. A. Lenard and E. J. Patzner, "A Survey of Crystal-Growing Processes and Equipment," *Semiconductor Products and Solid State Technology* 9, no. 8 (August 1966): 35–42.

Growth of Single Silicon Crystals: Resistance-Heated Crystal Puller

Early crystal pullers for the fabrication of silicon single crystals used radio frequency heaters to melt the silicon being processed. As experience with the process grew, it became clear that resistance heating was both more efficient than radio frequency heating and also more precisely controllable—a very important attribute. As crystal pullers were built that could produce larger diameter crystals, radio frequency heating was increasingly preferred and ultimately became the only type used in production.

Development History
In the mid-1950s it became clear that users preferred the resistance-heated pullers they were building in-house for use in the production of silicon single crystals. National Research Corporation became interested in producing a commercial resistance-heated puller because it was looking for products in which its knowledge of vacuum technology would prove useful. Joseph Wenkus, then an employee of Microwave Associates, a user of crystal pullers, was hired as a technical consultant on the project. Microwave Associates had earlier received a contract from the U.S. Army Electronics Command, Fort Monmouth, New Jersey, to manufacture microwave silicon diodes. As part of fulfilling this contract, Microwave Associates had built a resistance-heated silicon crystal puller about 1953 or 1954. Wenkus transferred the insights he had gained on this project to National Research Corporation, which developed a commercialized version of the equipment.

Commercialization
National Research Corporation introduced its resistance-heated silicon puller in the winter of 1957.

Growth of Single Silicon Crystals: Dislocation-Free Crystal Puller

Imperfections in the lattice structure of single crystals are called dislocations. Any reduction in the number of dislocations in the silicon crystals used to build semiconductors improves the properties of those semiconductors.

Development History
The original work on dislocation-free crystal pulling was done by William C. Dash at General Electric's Schenectady Research Laboratories. General Electric maintained a fairly strong program in materials-purification technology and was a leader in the development of silicon-powered diodes and silicon-controlled rectifiers. The thermal stability and lower resistance of the dislocation-free material was important to the construction of these devices. Dash began his work on dislocation-free growth of crystals about 1956. In April 1959 he published a procedure for the growth of dislocation-free crystals using existing crystal pullers.[1] The essence of the technique was to initially pull the crystal from the melt at a rapid rate that results in a small diameter rod of single crystal. This procedure allows any dislocation existing in the seed crystal to "grow out." After this result has been achieved, the pulling rate is slowed and the rod diameter is allowed to increase to the desired size.

Once Dash published the technique for growing dislocation-free crystals, most semiconductor manufacturers adopted it.

Commercialization
In 1968 Siltec introduced dislocation-free crystal pullers commercially. The hardware difference between these pullers and earlier pullers is relatively small. According to the manufacturer, the dislocation-free pullers required slightly greater standards of cleanliness and temperature control. Appropriate user technique, however, was still required to achieve the desired dislocation-free output.

Note
1. William C. Dash, "Growth of Silicon Crystals Free from Dislocations," *Journal of Applied Physics* 30, no. 4 (April 1959): 459–74.

Growth of Single Silicon Crystals: Automatic Diameter Control

The diameter of the single crystal rod produced by a crystal puller is critically dependent on the rate with which the crystal is pulled from the melt. Since standard diameter rods are needed for further production steps, any oversized rod sections must be ground down and undersized rods must be eliminated before further processing can take place. An automatic diameter control system effectively eliminates either over- or undersized rods, thus eliminating the associated production costs and waste.

Development History

The automatic diameter control system for crystal pullers is essentially an infrared optical system that observes the diameter of the rod being pulled and controls the rate of pull accordingly. Initial work on such systems began at IBM's East Fishkill, New York, facility about 1964 or 1965. The IBM effort was headed up by E. J. Patzner, R. G. Dessauer, and M. R. Poponiak. At that time IBM was in great need of single-crystal silicon for semiconductor production, having just entered into the production of the 360 Series computer. The firm looked to automatic diameter control as one means to increase production capacity.

An initial pilot system using automatic diameter control was in use in 1966. It was turned over to IBM's production department in early 1967. In October 1967, IBM researchers published a very detailed article describing the system and its capabilities.[1]

Commercialization

The first firm to manufacture the automatic diameter control system commercially was Hamco Corporation of Rochester, New York. Hamco introduced it as a feature on the first crystal grower of their manufacture, which they introduced in 1967. Hamco did not license the technique from IBM—they simply adopted it. Other firms did license the system from IBM at a fee that IBM describes as nominal. The first firm to license from IBM was National Research Corporation, which already made crystal growers. Indeed, the original system developed by IBM was built up on an NRC crystal puller.

Note

1. E. J. Patzner, R. G. Dessauer, and M. R. Poponiak, "Automatic Diameter Control of Czochralski Crystals," *Semiconductor Products and Solid State Technology* 10, no.10 (October 1967): 25–30.

Float Zone Crystal Growing

The float zone method of manufacturing single silicon crystals is an alternative to the dislocation-free crystal-pulling technique (which innovation see). It saw little commercial use in the United States but was extensively used in Germany—especially by Siemens. In the float zone technique, a rod of multicrystalline silicon is held in a vertical position by supports attached to both its ends. A small heated ring circles—but does not touch—one portion of the silicon rod. A seed crystal of silicon is next placed against one end of the rod and the heating ring is brought adjacent to it. The heating ring melts the silicon adjacent to the seed crystal. Slow movement of the ring down the length of the rod causes the molten silicon to crystallize into a single crystal following the pattern of the seed crystal. The surface tension of the molten silicon zone keeps the molten material in position in the crystalline silicon rod as the process is carried out.

Development History

Zone refining, a precursor of floating zone crystal growing, was developed by W. G. Pfann, a metallurgist at Bell Laboratories, in 1951. The Pfann process was similar to the floating zone process just described, but it involved placing the material to be crystallized in a crucible rather than suspending it in space. Since molten silicon reacts chemically with most container materials, the Pfann process did not prove usable with silicon. About 1952 H. C. Theurer, also at Bell Laboratories, conceived of the float zone process and reported it in his laboratory diary. Slightly later and independently, Amis and Siemens as well as Paul H. Kech of the U.S. Army Signal Corps developed a similar technique. Kech published his results in 1952.[1] Later, Theurer's diary showed his priority in the invention; eventually Bell Laboratories was assigned a patent.[2]

Commercialization

According to H. C. Theurer, the first firm to commercialize the float zone technique was the German firm of Siemens in the early 1950s. Siemens used the process in the course of manufacturing silicon semiconductors. The first firm to commercially manufacture a float zone process machine was Ecco Corporation, North Bergen, New Jersey, in 1953. Ecco was approached by Paul Kech of the Army Signal Corps, one of the developers of the float zone technique. Ecco was a manufacturer of high-frequency induction heating devices, and Kech placed an order with them for an induction heating device appropriate for the float zone refining of silicon material. After the sale of the initial device ordered by Kech, Ecco modified and enlarged the float zone device, and today it manufactures float zone refining equipment for commercial sale.

Notes

1. Paul H. Kech and Marcel J. E. Golay, "Crystallization of Silicon from a Floating Liquid Zone," *Physical Review* 89, no. 6 (15 March 1953): 1297.
2. U.S. Patent No. 3,060,123.

Wafer Slicing: OD Saw

After a rod of silicon single crystal is manufactured, it must be sliced into very thin circular wafers. The blade thickness of the saw has an important effect on the yield of wafers from a given length of a single crystal rod. Obviously, the thicker the blade, the greater the proportion of single crystal that will be reduced to dust. The so-called OD saw was apparently first used to slice the silicon single crystals used to manufacture semiconductors. The design of the saw is somewhat like a bread slicer with a rotary blade—the outer diameter of which (abbreviated as OD) bears against the single crystal to be sawn.

Development History

Little is known about the development of OD saws. Apparently they were used in a range of industries prior to their application to the semiconductor industry. Who first applied them to the task of slicing semiconductor wafers is not known.

Commercialization

It is not known who was first to commercially manufacture OD saws. Apparently a firm called Do-All was an important supplier of such machines to early semiconductor firm purchasers.

Wafer Slicing: ID Saw

The ID saw has a significantly thinner blade than that used on an OD saw (see preceding entry). The blade of the ID saw is a thin sheet of metal of circular shape with a large circular hole in the middle. This circular blade is clamped around its entire outside circumference, and the edge of the large hole in its center is used as a cutting blade—thus the name ID saw. Material to be sliced is slid into the hole, which is larger than the diameter of a silicon single crystal rod, and then pressed against the cutting edge for slicing.

The rigid clamping of an ID saw blade around its entire outer circumference and the application of some radial tension to the blade makes an ID saw blade much more rigid than an equally thick OD saw blade. Thus ID saw blades of similar performance can be thinner, increasing the number of wafers that can be sliced from a given rod of a silicon single crystal.

Development History
Little is known about the development history of the ID saw. We have learned of, but have been unable to trace, a British patent credited to Sayers that describes the ID slicer. We have also been told of an article in the field of dental research that appeared in a Belgian journal in the mid-1950s. This article reportedly describes an ID saw being used as a tooth slicer.

Commercialization
Capco Corporation, a British company, reportedly sold the first ID slicers used in the semiconductor industry to GE and Raytheon around 1960. In 1961 the Do-All Company apparently licensed the British patent and started production of ID saws. In that same year Hamco of Rochester, New York, began to produce ID saws without a license. Hamco's decision to proceed without a license was apparently based on recollection of the Belgian article, whose publication date preceded the date of the British patent.

Wafer Polishing: Silicon Dioxide Chemical/Mechanical Process

After wafers are sawn from single crystal rods of silicon, their surfaces must be polished. Original practice in the semiconductor industry was to polish these wafers much as glass lenses are polished in the optical industry, that is, with very fine abrasives. Polishing by abrasion was slow and did not leave an entirely damage-free wafer surface on completion. The silicon dioxide chemical/mechanical polishing process cut the polishing time from approximately 1½ hours to approximately 5 minutes. It also resulted in a damage-free surface.

Development History
The silicon dioxide chemical/mechanical process of polishing silicon wafers was developed by R. J. Walsh and A. Herzog of Monsanto Corporation.[1] Monsanto was in the business of supplying polished silicon wafers to the semiconductor industry and Walsh and Herzog were aware of the need for an improved process. The process developed by Walsh and Herzog was not developed theoretically, according to Walsh. When searching for better mechanical abrasives, Walsh experimented with submicron particles of silicon dioxide suspended in an alkaline fluid. The excellent results achieved

with this material turned out to be due to chemical as well as mechanical factors—but this result was serendipitous. Monsanto first sold silicon wafers that had been polished by the silicon dioxide chemical/mechanical technique in late 1962.

Commercialization

The silicon dioxide chemical/mechanical process for polishiing silicon wafers can be implemented by using the innovative polishing material in existing polishing machines. According to Monsanto, most suppliers of polished wafers are using the Monsanto technique, but only Western Electric has licensed it from Monsanto. Since no special-purpose machinery is required, no equipment supplier has become involved in the commercialization or dissemination of this innovation.

Note

1. R. J. Walsh and A. Herzog, U.S. Patent No. 3,170,273, issued 23 February 1965, and assigned to Monsanto. Also, Erich Mendel, "Polishing of Silicon," *Semiconductor Products and Solid State Technology* 10, no. 8 (August 1967): 36–37.

Wafer Polishing: Cupric Salt Chemical/Mechanical Process

The cupric salt chemical/mechanical method of polishing silicon wafers was significantly faster than the silicon dioxide chemical/mechanical process used previously (see preceding entry). The process also provides a damage-free mirror surface on silicon wafers.[1]

Development History

The cupric salt chemical/mechanical method of polishing silicon wafers was developed at IBM. IBM manufactures its own semiconductors on a large scale. Researchers were concerned that the technique that IBM was using left imperfect wafer surfaces. Three researchers at IBM had an idea that a solution of water-soluble salts of fluoride and nitride would produce a very slow etching reaction that would improve the quality of the surface. When they tried this solution of salts, they found that the wafers produced were, indeed, perfectly polished. After continued experimentation, however, they found that the valuable polishing action was actually being achieved by cupric ions. These ions were being introduced to the solution inadvertently through corrosion of some of the copper plumbing in the experimental equipment. When this serendipitously discovered effect was understood, they found they could control it well enough for use in commercial production.[2]

Commercialization

IBM began to use cupric salt chemical/mechanical wafer polishing on its in-house production of silicon semiconductor wafers in 1966. It is not clear whether any other firms are using the technique. The innovation was an innovation in polishing solutions only; no change in existing polishing equipment was required.

Notes

1. L. H. Blake and E. Mendel, "Chemical-Mechanical Polishing of Silicon," *Solid State Technology* 13, no. 1 (January 1970): 42–46.
2. G. A. Silvey, J. Regh, and Gardiner, U.S. Patent No. 3,436,259, assigned to

IBM. Also J. Regh and G. A. Silvey, paper presented at Electrochemical Society Meeting, Philadelphia, Penn., 14 October 1966.

Epitaxial Processing: Pancake Reactor

Epitaxy is an important technique used in the fabrication of some semiconductors. The process involves passing a gas containing atoms of silicon or other desired materials over the heated surface of semiconductor wafers. Under proper conditions, atoms contained in the vapor will attach themselves to the surface of the wafer in a manner that continues the crystalline structure of the underlying wafer. The end result is a layer of desired properties that has, in effect, been grown onto the surface of the silicon wafer.

Development History

Researchers at Bell Laboratories had investigated the growing of epitaxial layers on germanium as early as 1950. In 1960 they announced the process as a commercial reality.

Commercialization

Engineers at Western Electric entrusted with setting up a production process for semiconductors containing epitaxial layers began to search for a supplier of the needed equipment—a pancake epitaxial reactor. Ecco Corporation, North Bergen, New Jersey, was conveniently located and capable of performing the work. Ecco was given the design requirements—possibly, but not certainly, including blueprints. Ecco shipped the first commercial epitaxial reactor to Western Electric in September 1961.[1] Sales of Ecco Corporation in 1961 were approximately $1 million.

Note

1. V. Y. Doo and E. O. Ernst, "A Survey of Epitaxial Growth Processes and Equipment," *Semiconductor Products and Solid State Technology* 10, no. 10 (October 1967): 31–39.

Epitaxial Processing: Horizontal Reactor

The horizontal epitaxial reactor performs the same function as the pancake epitaxial reactor (see preceding entry) but is improved in so many respects that it represents a new generation of equipment. Horizontal reactors have an increased wafer capacity relative to pancake reactors and, because of increased automation, yield more uniform results.

Development History

Apparently numerous users experimented with horizontal epitaxial reactor configurations, one of these being Motorola, which in 1961 developed a production horizontal epitaxial system for its own use. Motorola applied this system first to germanium wafers and shortly thereafter to silicon.[1]

According to interviewees at Motorola, much of that firm's technological information related to horizontal epitaxial reactors came from publications that reported some English experimental work. Motorola took out some patents on aspects of their horizontal reactor about 1961. Names reportedly on the patent were T. Law, W. Corrigan, G. Russel, and Klink.

In late 1962 or early 1963 an engineer who had worked on the Motorola project, Larry Jo, left Motorola and began to work at Fairchild Semiconductor. The latter firm was working on an improved horizontal reactor; in 1964 Fairchild did develop both an improved reactor that had larger capacity than earlier reactors and a susceptor made of silicon carbide. (The susceptor is the surface upon which the wafers being treated lie. The wafers can be contaminated by the susceptor if improper materials are used.) Jo found that the Fairchild system was very operator dependent, that is, reproducibility of results was greatly dependent on the individual operating the system. He, therefore, set about developing an automated system.

Commercialization

The first firm to produce the horizontal epitaxial reactor commercially was Semi-Metals, a New Jersey firm in 1965. This firm had not previously been in the business of manufacturing equipment for the semiconductor industry, but was involved in the manufacture of silicon wafers. It occurred to that firm, however, that manufacture of a horizontal epitaxial reactor might be a commercially profitable venture. Larry Jo was hired from Fairchild to be the engineer in charge of the project. According to Jo, the reactor commercialized by Semi-Metals was effectively the same as the one developed by Fairchild in 1964.

Note

1. V. Y. Doo and E. O. Ernst, "A Survey of Epitaxial Growth Processes and Equipment," *Semiconductor Products and Solid State Technology* 10, no. 10 (October 1967): 31–39.

Epitaxial Processing: Barrel Reactor

The barrel epitaxial reactor serves the same function as the original pancake epitaxial reactor (which innovation see). However, it represents an improvement over both the pancake and horizontal reactors in terms of capacity as well as in terms of uniformity of results obtained. The barrel reactor gets its name from its geometry. Silicon wafers to be processed are attached to the outer surface of a cylinder looking somewhat like a barrel. This cylinder is then placed within another concentric cylinder, and the gas used to treat the wafers is passed between the two cylinders while the inner cylinder rotates. The rotation of the inner cylinder insures that all wafers are exposed to the same processing conditions.

Development History

Work on the barrel reactor was begun by IBM in the early 1960s. IBM was driven by a need for higher production capacity for epitaxial wafers. In 1965 the development was presented by IBM researchers at a meeting of the Electrochemical Society.[1] A patent was assigned to IBM in 1969. Apparently several other firms—including RCA, Texas Instruments, and Motorola—replicated the IBM barrel reactor for their own use in the 1960s. It is not clear whether or not these firms licensed the technology from IBM.

Commercialization

In 1971 the barrel epitaxial reactor was commercialized by Applied Materials Technology (AMT), Santa Clara, California. The AMT product utilized the basic configuration of the IBM system. However, it also included an AMT-developed radiant heating system rather than the radio frequency heating system used by IBM. (Wafers are heated as part of epitaxial processing.) AMT claims the use of radiant heat results in a pure and more uniform product and they have patented it.[2]

Applied Materials Technology had sales of less than $16 million in 1971, the year in which it commercialized the barrel reactor.

Notes

1. E. Ernst, D. Hurd, G. Seeley, and P. Olshefski, "High-Capacity Epitaxy Machines," paper presented at Electrochemical Society Meeting, Buffalo, N.Y., 10–14 October 1965.

2. V. Y. Doo and E. O. Ernst, "A Survey of Epitaxial Growth Processes and Equipment," *Semiconductor Products and Solid State Technology* 10, no. 10 (October 1967): 31–39.

Resist Coating: Wafer Spinner

One of the early steps in semiconductor manufacture involves coating of a wafer with a thin film of photoresist—a material formulated to change chemically when exposed to light. This change allows one to selectively remove either the exposed or unexposed portions of the resist film by further chemical processing. It is important to the manufacture of quality semiconductors that the resist coating be put on in a very thin, even layer. Historically, this has been accomplished by placing wafers on a disk, putting a drop of resist on the wafer surface, and then spinning the wafers very rapidly to achieve a thin, even film of resist. The machine that performs these functions is called a wafer spinner.

Development History

It is not clear which user firm was first to develop wafer spinners. It is clear, however, that user firms developed these spinners before commercial equipment producers did. In June 1962 employees then at Fairchild, a major semiconductor maker, report that spinners were in use. These were crude machines—simple aluminum plates with a lip on the outer diameter and a motor underneath. Operators would place several wafers against the lip of the rotor, drip resist on each, and then turn on the motor.

Commercialization

The first commercially produced resist spinner was introduced commercially during or before 1964. Industry participants no longer recall, however, which firm was first to commercialize a spinner. Candidate firms are Micro Tech Manufacturing, Transmask, Preco, and Applied Engineering.

Resist Coating: High Acceleration Wafer Spinner

Wafer spinners are used to spread thin coatings of resist on semiconductor wafer surfaces (see description of basic wafer spinner above). High-speed photography showed that most of the spreading of the resist occurred within the first few revolutions of the spinner and that high acceleration of the spinner could produce a thinner, more uniform coat of resist—both very desirable characteristics.

Development History

The high acceleration spinner was developed by Vern Shipman, an engineer who founded a firm called Head Way Research Corporation, Garland, Texas. Shipman had worked at Collins Radio exploring how to put thin film resist on quartz substrates. He started his firm in June 1964, and in October 1964 decided to build a resist spinner as his first product. At that time he did not have semiconductor manufacturers in mind as customers; instead, he was thinking of electronics firms (e.g., Collins Radio) that needed thin films in order to manufacture other kinds of electronic components. He completed his first machine in December 1964.

Commercialization

Shipman sold his first machine to Texas Instruments in December 1964. From the user's point of view, the high acceleration feature of the Head Way spinner was perhaps most important, but the machine had numerous other useful features as well such as automatic braking on the completion of the resist coating cycle; this increased its productivity.

The high acceleration wafer spinner was the first product of Head Way Research.

Mask Alignment: Split Field Optics

Several masks—each precisely aligned with preceding masks—are required in the fabrication of a typical integrated circuit semiconductor. The mask alignment procedure first used in the industry involved marking two dots at widely different points on a wafer along with two similarly positioned dots on each mask. Operators would then align succeeding masks by visually superimposing the dots on each mask with the dots on the wafer surface.

Early process equipment consisted of a microscope that would allow an operator to focus on one of these dots at a time. The operator would first align a dot on the wafer with a matching dot on the mask as best he could: He would then shift to the other dot and repeat the operation. Since alignment activities on the second dot would disturb the alignment on the first dot, several shiftings back and forth were required before the process was completed. Using this system, skilled operators might be able to align five or six masks per hour.

The introduction of split field optics allowed the operator to see both dots at once and to align both at once. This relatively simple innovation increased the productivity of operators performing mask alignment tenfold.

Development History

Users who developed split field optics and other process innovations typically preferred not to publicize their innovations: They wished to keep knowledge of their success from their competitors. We are thus forced to rely on the recollection of interviewees as to priority in the development of split field optics. According to industry interviewees, Jim Nall at Fairchild was the first to develop a split field alignment system in late 1959 or early 1960.

Commercialization

The first firm to offer split field optic alignment systems commercially was Micro Tech Manufacturing (now a division of Sprague Corporation) in 1963.[1] It is not clear whether Micro Tech developed the split field optics concept itself or whether it acquired it from a user. Ed Forcier, the founder of Micro Tech, was an employee of Fairchild at the time that Nall developed the split field system there. Thus, he might have been in a position to learn of it at Fairchild. On the other hand, Ed Jasiewicz, general manager of Micro Tech in 1974, felt that the system was developed entirely in-house.

In 1964 Kulicke and Soffa (K&S) also announced a split field mask alignment system. The genesis of their product was, they report, an approach by a representative of Nikon, the Japanese optical firm, who offered to sell them a set of split field optics usable in a split field optics mask alignment system.

Note

1. Micro Tech Manufacturing, Advertisement, *Semiconductor Products* 6, no. 8 (August 1963): 62.

Mask Alignment: Automated System

An automated mask alignment system automatically performs the alignment work previously performed by an operator using a split field optical system (see preceding entry). The principal advantage of the system is increased speed and increased accuracy—the latter improves the yield of finished semiconductors of good quality.

Development History

The key firm in the development of automated mask alignment systems was Computervision of Medford, Massachusetts. This firm is an equipment manufacturing firm; according to industry interviewees, no user development work preceded their efforts.

Computervision was formed in the late 1960s by three engineers, one from Singer Corporation and two from Concord Controls Corporation. Their goal was to produce automated devices. One of their initial products was an automated mask alignment system. According to Michael Cronin of Computervision, the company founders were told by most manufacturers that it was impossible to build an automated alignment system and that IBM had apparently tried with no success to fabricate one. Computervision initially tried to induce Kulicke and Soffa, a firm well-known for optical alignment systems, to engage in a joint venture. When Kulicke and Soffa expressed no interest, Computervision bought one of the standard K&S systems, modified it for automatic operation, and then showed it to Kulicke and Soffa. At last, K&S was sufficiently impressed with it to enter into a joint venture for the production of the device.

Commercialization

The first models of the K&S/Computervision auto aligners were offered at the WESCOM show in 1970. About 50 of the systems were sold, but they had problems that, a year or two later, led K&S to discontinue its involvement with auto aligners and with Computervision. Computervision then acquired Cobilt Corporation in California, a small concern producing manual aligners, and in 1972 introduced a new series of auto aligners, which became the first fully successful commercial product in the field.

Silicon Junction Fabrication: Diffused Junction Furnace

Silicon transistors are made up of regions "doped" with *p*-type and *n*-type impurities. Where these regions meet, a so-called junction is formed. Such junctions are the real operating heart of semiconductor devices.

The first silicon transistors sold commercially had so-called grown junctions. Grown junction technology appears to have been developed independently at Bell Laboratories and at General Electric. At Bell Labs the initial work was carried on by Morgan Sparks and Gordon Teal, with the first publication regarding the process appearing in 1951.[1] Work at GE on grown junctions was carried out by R. N. Hall and his associates. Grown junctions are created by simply adding impurities of the proper type to a silicon crystal as it is being grown. For example, if one begins to grow a crystal from molten silicon containing a *p*-type impurity, then the resulting crystal will be *p*-type. Addition of sufficient *n*-type impurity to overwhelm the *p*-type impurity during the growing process of this crystal will result in a switch from *p*-type to *n*-type silicon—with the point at which the transition takes place being a grown junction.

The dominant method used to produce silicon junctions today is called the diffusion method. This method was developed at Bell Laboratories, which called a second symposium of its licensees to inform them of the technology in 1956. This process involves heating silicon wafers in a furnace to a precise temperature for a precise time, and at the same time passing a gas containing the desired *p*- or *n*-type material that one wants to use to form a junction through the furnace. The high temperature in the furnace allows atoms of the impurity to diffuse into the crystal structure of the silicon to a known concentration and a known depth, thus creating a diffused junction. The process equipment used to create diffused junctions is called a diffusion furnace. It is essentially a furnace in which the temperature and the introduction of the impurities can be controlled very precisely.

Development History

The first diffusion furnaces used to manufacture diffused junction transistors commercially were built by user firms. Later, manufacturers of laboratory furnaces custom-built furnaces to user specifications.

It took several years for the furnace users to fully understand what the critical parameters of the diffusion process were. When it became apparent that the temperature stability and control were extremely critical to the production of good diffused junction semiconductors, furnace manufacturers were able to respond very quickly with precisely controllable furnaces.

Commercialization

In 1961 the Heavy Duty Electric Division of Sola Basic Products Company introduced a line of special-purpose diffusion furnaces at the WESCOM show.

Note

1. W. Shockley, M. Sparks, and G. K. Teal, "*p-n* Junction Transistors," *Physical Review* 83, no. 1 (1 July 1951): 151–62.

Silicon Junction Fabrication: Ion Implantation Accelerator

Silicon junctions in semiconductors are created by bringing *n*-type material into close contact with *p*-type material. This is done by introducing *n*-type impurities into a region of a *p*-type material, or vice versa. Ion implantation—a method of introducing such impurities—has the potential of being more precise than the thermal diffusion method (see preceding entry). The method involves first ionizing atoms of the desired impurity and then accelerating them to a known speed by means of an ion accelerator. The ion beam thus created is aimed at specified points on the silicon wafer surface, and the ions penetrate into that surface to a precisely known depth and concentration.

Development History

Some early experimental work on ion implantation was performed at Bell Laboratories in the early 1950s.[1] The commercial success of the diffusion process, however, delayed serious commercial interest in the ion implantation technique until the 1960s. In the early 1960s a significant amount of experimentation with ion implantation was conducted by Hughes Laboratory at Newport Beach, California (H. G. Dill and R. W. Bowers were among the principal investigators); by Bell Laboratories at Murray Hill (A. McRae was a principal investigator); at AFCRL (Dr. Rooslid was a principal investigator); and at an Ion Physics joint venture with Signetics Corporation. The Ion Physics/Signetics group apparently did produce some ion-implanted devices for the U.S. Air Force in the early 1960s, and they published a discussion of methods for producing such devices in 1965.

Commercialization

Early experimental work on ion implantation was conducted using laboratory-type ion accelerators that were commercially available from several sources. Ion accelerators specifically designed to be used in the production of semiconductors were probably first commercialized by High Voltage Engineering in about 1969.

High Voltage Engineering was a partner in a venture to develop ion-implanted devices called Ion Physics/Signetics. In its work it did develop ion implantation accelerators, but it was prevented by its partnership agreement from selling these commercially—the venture was interested in selling devices only. When the Ion Physics/Signetics venture was dissolved in 1968 or 1969, High Voltage Engineering was free to pursue the sale of ion implantation accelerators. It sold only a few of these, however, before deciding that the field was commercially unattractive.

The second firm to commercialize an ion implantation accelerator specifically designed for semiconductor implantation work was Accelerators, Inc. Hughes had bought 20% of Accelerators stock in 1968 or 1969, and Hughes scientists designed the production-oriented accelerator sold by them. In 1970 Accelerators shipped their first production-oriented accelerator to Hughes at Newport Beach. Previously, Accelerators, Inc., had produced laboratory-type accelerators.

In 1971 Extrion Corporation began to produce production-oriented ion accelerators. Extrion was a start-up company formed by employees of High Voltage Engineering who left that firm when it decided to stop producing accelerators for semiconductor ion implantation work.

Note

1. W. Shockley, U.S. Patent No. 2,787,564, issued in 1954.

Scribing and Dicing: Automated Mechanical Device

In the process of making semiconductors, many circuits are fabricated simultaneously on a single wafer of silicon crystal. When completed, these circuits must be physically separated from each other, and this is done by breaking the silicon wafer on which they were fabricated into tiny square chips—each containing only one circuit.

In the earliest days of the industry, the scribing-and-dicing process step was carried out by a worker who, with the aid of jigs and fixtures, scratched lines manually between the circuits on a wafer with a sharp diamond point and then broke the wafer along those lines with the aid of a straightedge. This manual process was slow and subject to worker error. For example, a worker might easily tire and scribe a line through the middle of some circuits instead of along their edges—and thus ruin them. To improve yield and to lower costs, automatic machinery was developed to perform the scribing-and-dicing task. The initial generation of such equipment simply automated the process previously carried out manually. That is, a diamond point was mounted on a machine that would move the point in such a way as to scribe the wafers at the proper locations.

Development History

Several semiconductor manufacturing firms may have developed their own automatic scribing-and-dicing machines. The device with the closest link to the first autoscriber produced commercially was that built by Western Electric engineers at their Allentown, Pennsylvania, plant. Western began using their internally developed autoscribers in 1958 or 1959.

Commercialization

The first automatic scribing-and-dicing machine to be placed in the commercial marketplace was introduced by Kulicke and Soffa in March 1960. K&S was formed in the early 1950s as a machine design firm. In 1958 they observed the automatic scriber and dicer developed by Western Electric when they were invited into a Western Electric plant to perform an unrelated machine design task. K&S felt that there might be a commercial market for such a machine and proceeded to design a commercial version.

In 1965 Tempress also introduced a commercial automatic scribing-and-dicing machine. Tempress was formed by former Fairchild employees and, according to interviewees at that firm, the Tempress product embodied scribing-and-dicing technology developed at Fairchild.

Scribing and Dicing: Laser Scriber and Dicer

Laser scribing-and-dicing equipment performs precisely the same function in the semiconductor fabrication process as does the earlier mechanical scriber and dicer (see preceding entry). Its main difference is that it substitutes a laser beam for the diamond cutting tool used in the mechanical scribing-and-dicing machine. The laser beam offers the user two major advantages over the earlier diamond scribing machine. First, it is much faster. Second, the laser beam cuts much more deeply into the wafer than did the diamond cutting tool. These deep cuts make the process of separating the wafer into individual chips more accurate and thus increase yield.

Development History

Quantronics of Long Island, New York, was a manufacturer of lasers. In 1968 it was actively involved in trying to find applications for these. During a trip to Texas Instruments, a semiconductor manufacturer, Quantronics discovered that the firm had tried unsuccessfully to build a laser scriber system. They tried to persuade Texas Instruments that Quantronics could succeed where Texas Instruments had failed and that the latter should join them in a joint venture. Texas Instruments was not interested, however. Later in 1968, Quantronics was able to interest Motorola in a laser scribing system.

In 1968 Quantronics and Motorola entered into an informal joint effort in which Motorola supplied Quantronics with some development funds and a great deal of technical information on the desired characteristics of a laser scribing system.

Commercialization
Quantronics offered a laser scriber and dicer commercially in 1970. The first unit was shipped to Motorola. Immediately thereafter Texas Instruments purchased 24 units at $70,000 per unit.

In about 1971 Electroglas also commercialized a laser scribing-and-dicing machine. In the 1970s Quantronics and Electroglas held the major share of the market for laser scribing-and-dicing machinery.

Wire Bonding: Thermocompression Bonding

Semiconductor chips must be linked electrically to the outside world in order to function. This electrical linking is achieved by means of tiny metal wires that are physically bonded to the semiconductor chip on one end and to an electrical connector on the other. The bonding of the wire to the surface of a semiconductor chip is a difficult task. The wire must make a good electrical connection that will not degrade with time.

The earliest means of bonding wires to commercially produced silicon semiconductor devices involved tiny balls of solder that in essence soldered the wire to the chip. These solder bonds proved difficult to produce reliably. Sometimes the joint was mechanically weak; at others the solder joints would form a diode junction with the semiconductor materials and thus degrade the performance of the semiconductor.

Thermocompression bonding, the innovation under consideration, involves heating the semiconductor surface to about 200°–300°C and then simply pressing the wire to be bonded against the semiconductor chip surface at the appropriate place with a pressure from 5000 to 10,000 lb/sq in. In a few seconds a bond with excellent physical and electrical properties is formed. The machine that performs this task is called a thermocompression bonder.

Development History
Thermocompression bonding was developed by three scientists of Bell Laboratories: O. L. Anderson, H. Christensen, and P. Andreatch. Most of the work of this team was performed in the 1955–58 time period. The team published its findings,[1] and the need was so great that many semiconductor manufacturers built their own thermocompression bonders in-house in the 1957–59 period.

Commercialization
The first commercial thermocompression bonder was offered by Kulicke and Soffa in late 1959. Demand for the product was so strong that within a year of introduction, net sales exceeded $1 million.

K&S observed thermocompression bonding in production equipment that was operating at a Western Electric semiconductor plant in 1958. K&S engineers made only minor mechanical modifications to the Western Electric thermocompression bonder before commercializing it.

Note

1. O. L. Anderson, H. Christensen, and P. Andreatch, "Technique for Connecting Electrical Leads to Semiconductors," *Journal of Applied Physics* 28, no. 8 (August 1957): 923.

Wire Bonding: Ultrasonic Bonding

The ultrasonic bonding of metals involves rubbing together the two pieces of metal to be bonded with such high energy that surface impurities on the two metal surfaces to be bonded are scrubbed away and the underlying atoms of metal brought into close enough contact to form a good bond. In ultrasonic bonding, the energy for mechanically rubbing the two pieces of metal together is provided by a tool that vibrates mechanically at an ultrasonic frequency.

Development History

Ultrasonic bonding as a general welding technique was discovered in the 1950s. Its first application to the attachment of wires to semiconductor chips apparently occurred in the mid-1960s, according to interviewees at Sonobond Corporation, Westchester, Pennsylvania, a supplier of ultrasonic welding equipment. Who was actually first to develop an ultrasonic bonder for semiconductor manufacturing use is not clear. It appears that Sonobond initially supplied ultrasonic transducers—the generators of ultrasonic energy needed for welding—to Fairchild and Motorola who then designed the first ultrasonic bonding equipment in-house.

Commercialization

Ultrasonic bonding equipment for the bonding of wires to semiconductor chips was commercialized by Sonobond in the 1960s. The corporation initially provided only the source of ultrasonic energy to semiconductor manufacturers but later commercialized a complete machine for the ultrasonic bonding of wires to semiconductor chips.

Mask Graphics: Optical Pattern Generator

The patterns of light and shadow that expose the resist-coated semiconductor wafers are generated by so-called masks. These masks must be very precise, and their creation is a demanding task. They are created on a large scale—perhaps 5 ft by 5 ft—and then photographically reduced. Masks were first created by a hand technique much like drafting: A desired pattern was cut out of a material called Rubylith by means of a straightedge and a small hand-held knife. When integrated circuits became more complex, the patterns needed in masks became more complex, and it was no longer practical to produce these masks by hand. The solution to the problem was a numerically controlled machine that exposed a large sheet of photographic material in the desired pattern.

Development History

Interest in automated mask generation began in the mid-1960s with the realization that large-scale integration required masks too complex to be cut manually. Bell Laboratories and IBM were among the first to attempt to create automated mask generation machines. Their initial attempts involved computer-driven knives that basically automated the former handwork of cutting a Rubylith pattern.[1] This technique was eventually found unpromising: Even though the Rubylith material was automatically cut, human operators still had to select which sections of material were to be stripped away from the pattern and which left behind. Mistakes made in this process were as devastating as mistakes made in the cutting process.

The innovation developed at IBM and elsewhere involved mounting a large photographic plate on a table that could be moved in the *x* and *y* directions under numerical control. A photographic projector would then project various-sized rectangles as specified by a computer program in order to create the pattern of light and shadow specified for a particular mask design.[2]

Commercialization

In July 1967 R. C. Beeh, an employee of OPTO Mechanisms, described a device that appeared to be a commercial automated pattern generator.[3] OPTO Mechanisms is, however, no longer in business and, although the product pictured in Beeh's article appears to be a commercial device, it is not clear that one was ever sold. (According to industry interviewees, OPTO Mechanisms did receive an order for one unit from Texas Instruments. That unit was never accepted by Texas Instruments, however, which apparently contributed to OPTO Mechanisms' eventual bankruptcy in about 1971.)

The first automated optical generator for the manufacture of semiconductor masks that was produced by an equipment manufacturer and successfully sold was put on the market in April 1968 by GCA/David W. Mann Company. The Mann system was much like the IBM system described in July 1967. The machine involved a numerically controlled pair of perpendicularly mounted slits. A light was mounted behind the slits and, depending on the position of the slits, was projected on a photographic plate mounted on an *x-y* table as a rectangle of any desired dimension. A suitable choice of rectangles and suitable positionings of the *x-y* table could be combined to build up any desired mask pattern.

Since its introduction in 1968, the original model 1600 optical pattern generator has been steadily improved. By 1976, a model was available that was about 10 times faster than the original model.

Notes

1. See H. O. Hook, "Automated Mask Production for Semiconductor Technology," *Semiconductor Products and Solid State Technology* 10, no.7 (July 1967): 35–38; and P. Donald Payne, "Photomask Technology in Integrated Circuits," *Semiconductor Products and Solid State Technology* 10, no.7 (July 1967): 39–42.
2. See note 1.
3. Roland C. M. Beeh, "A High Accuracy Automated Microflash Camera," *Semiconductor Products and Solid State Technology* 10, no.7 (July 1967): 43–49.

Mask Graphics: Electron Beam Pattern Generator

The major advantage electron beam mask pattern generation devices provide relative to optical pattern generators is their higher resolution: Optical systems are limited to about 7 Å spot size, whereas electron beam systems can produce a ½ Å spot size. The higher resolution obtainable with electron beam devices in turn allows users of the systems to make more complex and denser masks for integrated circuits.

Development History

The early history of electron beam pattern generators is difficult for us to discover because a great deal of the early work was apparently done outside the United States. The earliest advertisement we can find for a commercialized electron beam device appears in *Solid State Technology* in 1968: It describes a computer-controlled masked generator of the electron beam type offered for sale by JEOL, a Japanese corporation. The first work in the United States on an electron beam mask generator was apparently begun in 1966. In that year research was begun at a Bell Laboratories research center in Princeton, New Jersey, that resulted in the creation of experimental devices during the next three or four years. In 1971 Bell Laboratories used a system of its own design to manufacture actual integrated circuit masks at its Murray Hill facility.

Commercialization

The first electron beam mask pattern generator offered for commercial sale in the United States was apparently that advertised by JEOL in 1968. The development history of that system and its eventual commercial success or failure are not known to us.

Mask Reduction: Two-Stage Step and Repeat Reduction Process

A two-step process is used to reduce the black-and-white pattern for an integrated circuit mask from its original size of perhaps several feet square to the miniature size necessary for actual use in connection with producing tiny integrated circuits. The process consists first of reducing the full-size layout by a factor of perhaps 20 by means of a precision camera. A second precision camera then reduces the image by another factor of 20 or so and then "steps" the reduced image across a photographic plate, thus creating multiple images of the same mask pattern. This plate is then used to produce multiple integrated circuits on a single wafer of silicon.

Development History

The two-stage step and repeat mask reduction process was developed almost simultaneously by two user groups and by a manufacturer of equipment—all apparently working independently. The first work was done by users at Fairchild Semiconductor about 1959. Gordon Moore of Fairchild recalls that Bob Noyce constructed a two-step photo-reduction system for internal use at Fairchild about 1959. Fairchild, however, made no attempt to commercialize this device; indeed, it strove earnestly to keep it secret.

From 1959 through 1961 researchers at Diamond Ordnance Fuze Laboratories, located in Washington, D.C., also developed a two-step mask reduction process. T. C. Hellmers, Jr., and J. R. Wall reported their work in detail in a January 1961 article.[1] About 1960 GCA/David W. Mann Company began to develop a two-stage step and repeat mask reduction machine for commercial sale. That firm had been receiving orders for masks from semiconductor firms (the first such firm being Transitron) since early 1960. Since GCA/David W. Mann was essentially an instrument-manufacturing company, not a service-providing company, their thoughts naturally turned to manufacturing a device rather than providing a service.

Commercialization

GCA/David W. Mann was the first firm to make a two-stage step and repeat mask reduction device available to semiconductor manufacturers commercially. They first showed the device in the spring of 1961 and sold the first one to Clevite Corporation on 28 April 1961.

GCA/David W. Mann had sales of under $2 million in 1961. Fairchild Semiconductor had sales of approximately $43 million in 1959.

Note

1. T. C. Hellmers, Jr., and J. R. Nall, "Microphotographs for Electronics," *Semiconductor Products* 4, no. 1 (January 1961): 37–42.

DATA SET FOR PULTRUSION PROCESS MACHINERY INNOVATIONS

The innovations described in this data set are associated with pultrusion process machinery. The innovation history of the pultrusion process is followed from the original rudimentary equipment through the major process machinery improvements that have been commercialized over the years. Sample selection criteria are discussed in chapter 3.

Because the manufacturers and users of pultrusion equipment do not often document their innovations in publications, most of the information comes from interviews held with employees of pultrusion equipment manufacturer and user firms who had direct knowledge regarding the sampled innovations.

Original Batch Pultrusion Process

Pultrusion machines fabricate fiber-reinforced plastics products of constant cross-section. An everyday example of a pultruded product is the fiberglass-reinforced rod used by the makers of fiberglass fishing rods. In essence, pultrusion machines create pultruded products by quite a simple process. Reinforcing material such as fiberglass roving or cloth is first wetted with a thermoset resin such as polyester, then pulled through a heated block of steel (a die) that has a hole in the middle of a cross-section matching that of the desired pultruded product. When the wetted reinforcement has been drawn completely through the die, the resin has cured around and between the strands of reinforcing material, and the reinforced plastic product—a pultrusion—is complete.

Reinforced plastic product made by means of the pultrusion process is characterized by a very high percentage of reinforcing material that is aligned in known directions. (In contrast, ordinary reinforced plastic is characterized by short lengths of reinforcement fiber oriented randomly within the plastic matrix.) These characteristics make pultrusion especially appropriate for applications in which high stress is applied to the part in directions parallel to the embedded material.

Development History

The first pultruded products were long, thin fiberglass-reinforced cylinders used in the manufacture of fiberglass fishing rods. These were manufactured by Ocean City Manufacturing Company, Philadelphia, starting approximately in 1947. The pultrusion process developed by this firm was a simple batch process that involved hand tooling. Fiberglass rovings (a threadlike form of fiberglass reinforcement) were collected by hand into bundles about the thickness and length of a fiberglass fishing rod blank. These bundles were dipped into a container of room-temperature-cure polyester resin and then passed through a steel plate with a hole in it the size of the diameter of the rod being produced. This primitive die served to squeeze the rovings tightly together and eliminate excess resin. The resulting bundle was then hung from a hook until it cured.

Commercialization
The simple tooling required by this early pultrusion process was never produced commercially by a machinery builder. Today, this process is obsolete.

Intermittent Pultrusion Process

The first pultrusion process simply pulled a bundle of resin-wetted reinforcement through a die to shape it; it then would require several more hours to fully cure. If one could cure the pultrusion fully while it was still inside the die, one would have much more rapid processing times as well as more precisely dimensioned pultruded parts. The innovation described here was designed to obtain these advantages. Instead of simply being pulled through a thin plate for shaping as in previous practice, wetted roving was pulled into a tubular die several feet in length. Because this die was heated, the length of material within it was fully cured in a matter of minutes. After curing, pull was resumed on the now-cured section of pultrusion in the same direction as before. This pull drew the cured section out of the die and at the same time drew a new uncured wetted roving into the die behind it. When the die was once again fully filled with uncured material, pulling would stop and the curing cycle would begin again. A crucial (patented) invention that made the process possible involved placing a cooled section at the front of the heated die just described. This prevented resin from curing before it was pulled inside the die and formed to proper dimension.

Development History
The intermittent pultrusion process was developed by Roger White, president of Glastic Corporation, Cleveland, Ohio, in 1948. White had received a large order for a part of constant cross-section (spacer sticks for motor armature windings). As White recalls, attempts to fabricate the parts by conventional sawing and machining operations proved too expensive, thus inducing him to develop the new method.

Glastic Corporation was a small firm and did not have an established R & D budget. White estimates that he spent perhaps $5000 in direct costs to develop the intermittent pultrusion process.

Commercialization
White used the intermittent pultrusion process commercially within the Glastic Corporation from about 1950 to 1966. He tried to keep the process proprietary because it was significantly cheaper than alternative ways of making motor armature spacer sticks and allowed him to make a 30% or 40% profit on his sales of approximately $100,000 per year for the years 1951 through 1956. In 1956 a competitor hired away one of White's engineers and induced him to replicate White's proprietary process. As a consequence, profits for spacer sticks fell. White sued in an attempt to enforce his patent and after the expenditure of much time and money was awarded only $15,000 by the court, which also allowed the competitor to continue his infringement.

It appears that White's process machinery innovation was never used beyond the two firms discussed here. By the 1960s the continuous pultrusion process using a heated die (which innovation see) had advanced so far that Glastic discontinued the use of the process described here.

Tunnel Oven Cure

Some pultrusions need not be finished to a high tolerance (e.g., fishing rod blanks are machined after being pultruded; thus tolerances in the pultruded part are not critical). For such applications it would be advantageous to achieve the rapid processing rates characteristic of heat-cure pultrusion without the constraints of a heated die. Tunnel oven curing meets these requirements. In this process, fiberglass roving is wetted with resin, shaped, and then pulled through a tunnel oven—a long heated chamber in which the shaped roving and resin bundle is cured by radiant energy from the chamber walls rather than by contact with the heated die surface.

Development History

The tunnel oven cure/continuous pull pultrusion process was developed by Sam Shobert, president of the Polygon Corporation, Walkerton, Indiana, in 1950. At that time Shobert was involved in the manufacture of fishing rod blanks, using a slow, labor-intensive molding technique. Large customer orders prompted him to search for a faster method for making the product. Shobert estimates that the direct cost of the development project was less than $10,000 for materials, plus labor.

Commercialization

The tunnel oven cure/continuous pull pultrusion process worked well. Shobert's firm made hundreds of thousands of fishing rod blanks, and Shobert estimates that at "one time" they had about 80% of the market. Their profit on these was not high, however, ranging from 3% to 7% pretax. General Electric expressed some interest in using the process to make winding pins—a component of electric motors. Polygon gave them a license and built them two machines for a total price of approximately $50,000. Two years later, GE asked Polygon to take back the machines and produce the part for them, finding this a more economical arrangement. Polygon agreed, and produced $750,000 to $1,000,000 of the parts for GE for many years at a pretax profit of 20%.

In recent years Polygon discontinued the tunnel oven process in favor of standard pultrusion, using a heated steel die, which had improved over the years and had become a more economical alternative.

Heated Die Cure/Continuous Pultrusion Process

Prior to the die cure/continuous pultrusion innovation, pultrusions cured within a heated steel die could not be moved while curing. This was because polyester resin, the thermoset resin universally used in the early days of the pultrusion process, turned into an excellent adhesive when partially cured, gripping the die tightly at that point. Later, when fully cured, this stickiness disappeared and the completed pultrusion could be slid from the die. It was found that the mixing of lubricants such as carnuba wax into polyester prior to curing would prevent the partially cured polyester from sticking to the die surface. With this discovery, it became possible to create pultrusion process machines that steadily and continuously drew material through the die during the curing process.

Development History

The discovery of the value of internal lubricants in the pultrusion process and the development of the first truly continuous pultrusion process machine were achievements of Roy Boggs, an employee of Universal Molded Products Corporation in 1955 (today known as Morrison Molded Fiberglass Company, Bristol, Virginia). In 1955 Morrison Molded Fiberglass had obtained a large order for fiberglass handrails from the Federal Aviation Administration, and Boggs decided to experiment with new methods for producing these. No formal R & D budget was involved. He simply put equipment together out of material available on the shop floor. After numerous failures caused by partially cured polyester resin adhering to the inside of his steel dies, Boggs hit on the key idea of embodying lubrication in the resin itself. When he got the process to the point where it would run a few feet without jamming, he showed it to his management, which encouraged him to patent. Eventually Boggs obtained and assigned to Universal Molded Products more than 30 patents on the pultrusion process.

Commercialization

Universal Molded Products quickly became the largest manufacturer of pultruded products in the world. Its successor firm, Morrison Molded Fiberglass, retains this distinction today, having an estimated 20% market share.

Morrison Molded Fiberglass has not licensed its patents to U.S. competitors nor attempted to enforce its patent rights to the pultrusion process in the United States. It did not have a really basic patent and perhaps felt it could not prevail in U.S. courts. On the other hand, in 1960 it did begin an active program of licensing foreigners to use the pultrusion process and did and does build machines for foreign licensees. Foreign licenseees who buy pultrusion machines from Morrison pay an initial royalty fee of $40,000; then, they are subject to a contract that provides for continuing royalties. (When and if foreign licensees refuse to honor these contracts—as many have—Morrison does not attempt to enforce its rights legally, suspecting that likelihood of success would be slim.) Morrison's annual fees from foreign licensees from 1960 to present was not more than $100,000 per year. Morrison currently has 13 licensees.

The continuous pultrusion process reduced labor content from 50%/60% to 5%/10% of the product direct cost relative to the hand layup process it superceded. It is the basis of all current pultrusion processes.

Tractor Pullers

Initial continuous-pull pultruders supplied the mechanical force needed to pull the material through the process either by winches or by so-called nip roll pullers. Nip roll pullers consisted simply of powered wheels squeezing the pultrusion between them and pulling it along—much as the opposed rollers of an old-style washing machine laundry ringer drew material between them. Both systems had major drawbacks, however. Winches were obviously not suitable for providing continuous pull over a long span, and it was quickly found that nip roll pullers could not exert sufficient force. The traction required to pull pultrusions through a system was found to be as much as 200 lb/in. of perimeter of the pultrusion. To meet this problem, so-called tractor pullers were designed. These looked much like the treads of two bulldozers put face to face. They supplied greatly increased pulling power.

Development History

The first tractor puller used in pultrusion was designed by Roy Boggs of Universal Molded Products (currently Morrison Molded Fiberglass, Bristol, Virginia). The design task was found to be relatively difficult: Too much pressure and the pultrusion would be crushed; too little, and the tractor pullers would slip ineffectively. Boggs recalls that it took $20,000 in direct cost to build the first pullers capable of pulling pultrusions with a 14-in. perimeter.

Commercialization

The ineffectiveness of nip roll pullers for pultruding products with perimeters larger than 1 in.-or-so was glaringly apparent to all users of pultrusion process machinery. All these firms quickly turned to tractor pullers and made these standard industry practice. Since building pullers was relatively complex, and since Morrison—the innovator—would not build them for U.S. competitors, many users searched for outside suppliers. The firm that eventually emerged as a major producer of this component of pultrusion machines is the Gatto Corporation. This firm had earlier been in the business of tractor pullers for less demanding applications. When pultruders informed them of the new application, they proceeded to develop an appropriate puller. Gatto estimates that it cost them perhaps $50,000 to $100,000 to develop the new product.

Cut-off Saw

As the pultrusion process became more rapid and reliable, it became important to have some means of automatically cutting the continuously emerging pultrusion into the lengths required by customers. An automatically activated saw called a cut-off saw filled this need.

Development History

Automatic cut-off saws were used in numerous other industries for sectioning continuously produced products. It therefore seemed logical to apply this device to a similar task in the pultrusion industry. The adaptation, although very useful, seemed so unremarkable to participants in the industry that it is not clear to them who in fact was first to do so. Probably several users brought this innovation to the pultrusion industry independently.

Commercialization

The use of cut-off saws is universal in the pultrusion industry today. Since cut-off saws did not have to be specially adapted to the requirements of the pultrusion industry, no special source of saws can be pointed to. At the time cut-off saws were adapted to the pultrusion process, the price of such a saw was approximately $1500. Interviewees estimate it would have cost perhaps $500 to install the saw on a pultruder.

Augmented Radio Frequency Cure

Preheating of thick-walled pultrusions with radio frequency energy before they enter a standard steel die is the essence of the radio frequency augmented cure technique. It very effectively serves the purpose of accelerating the rate at which such products can be cured because heated steel dies can only heat a product from the perimeter, and it takes a long time for heat to penetrate to the center of a thick section. In contrast, radio frequency energy heats the entire product at once.

Development History

Radio frequency augmented cure was developed by Brant Goldsworthy in 1968. The innovation is simply implemented by placing a radio frequency heating unit ahead of the die location on an otherwise standard pultruder. At the time of the innovation,

Goldsworthy owned both a pultrusion process machine user company, Glastrusions, Inc., and a commercial manufacturer of pultrusion equipment, Goldworthy Engineering.

Commercialization
From 1968 to 1977 Goldsworthy sold perhaps a dozen radio frequency augmented cure systems. It was his practice to sell them as accessories, with a list price of approximately $25,000 for the commercial pultruders sold by Goldsworthy Engineering. No other firm has undertaken to supply such units to pultruders.

Preforming Tooling

As the size of pultruded products increased and as their shapes became more complex, it was no longer sufficient to simply pull the various strands of reinforcement material making up a pultrusion into the die. Instead, accurate placement of the various strands of roving and strips of reinforcement mat (a form of cloth made of reinforcement) became necessary. The range of racks and hooks and other specially designed tooling that performs this guidance function is called preforming tooling.

Development History
Preforming tooling is very much like software on computers in the sense that it must be designed anew for each specific application. Thus, every user who must pultrude a new product—or pultrude an old product on a new machine—is forced to design preforming tooling. The use of preforming tooling is now universal in the industry, except for the simplest shapes.

Commercialization
There are no commercial manufacturers of preforming tooling. All users devise their own as part of their product-engineering task.

Hollow Product Tooling

The development of hollow products—especially those with hoop strength—required tooling more complex than that required in the fabrication of solid products. Various solutions were developed, including apparatus for braiding and apparatus for winding strands of material in a hooplike fashion around the strands of reinforcement material being pulled through the process machinery.

Development History
Any of three pioneering pultrusion user firms may have been the first to develop hollow product tooling. Each used a somewhat different approach. Goldsworthy of Glastrusions used an overwinder approach; Shobert of Polygon used a braiding approach; and Boggs of Universal Molded products developed a very elaborate special-purpose approach. (Universal Molded Products had a massive order for 5-in. rocket tubes from the U.S. Army Chemical Corps. Approximately $700,000 was expended to tool up for this particular hollow product.)

Commercialization
There still is no standard approach in the field to manufacturing hollow products. User

firms build their own hollow product tooling for each product at a cost typically ranging from a few hundred to a few thousand dollars.

Improved Dies

Dies are built with a hole through the center that has the shape of the cross-section of a pultrusion being produced. For this reason a new die has to be manufactured for each new product one wishes to pultrude. Over the years, dies have been improved in several ways. They have been made smoother; they have been heated in a more precisely controlled manner; and their geometry has been improved. The result of these incremental improvements has been twofold: (1) the amount of scrapped product has been reduced down to approximately 5% from an initial rate several times higher and (2) the speed of the pultrusion process has been increased.

Development History
As noted above, dies are built anew for each product to be pultruded. Typically, they cost a few thousand dollars to build and their cost is charged to the customer as a set-up or tooling charge. The incremental improvements that have been applied to dies over the years are not the product of a separate R & D effort. Instead, lessons learned from previous die performance cause designers to modify new dies slightly. Modifications that result in improved performance are noted and become part of a firm's process know-how.

Commercialization
Owing to the special-purpose design of each pultrusion die, no firm manufactures them as a standard, commercial product. Dies are either built by user firms themselves or by tool and die firms to the specifications of user firms.

DATA SET FOR THE TRACTOR SHOVEL

The data set for the tractor shovel documents the development of the wheeled tractor shovel in 1939 and the 10 major innovations that significantly improved it over the succeeding 20 years. Sample selection criteria are found in chapter 3.

Because manufacturers and users of construction equipment do not often document their innovations in publications, most of the information contained in this data set comes from interviews held with employees of tractor shovel manufacturer and user firms who had been associated with the innovations studied. Interviewee recollections of dates of innovation commercialization have been checked against the dates of related advertisements that appeared in construction equipment trade journals.

The Original Tractor Shovel

The tractor shovel is a mobile, rubber-tired machine used for excavating and for the general handling of such bulk materials as earth, coal, and chemicals. A tractor shovel looks somewhat like a farm tractor with a large movable scoop mounted at the front end. Today, tractor shovels are produced in a wide range of sizes. Very large tractor shovels with massive 20 cu-yd scoops can be found working in open-pit mines, filling an entire truck with coal or ore with a single scoop. At the other extreme, one can find small tractor shovels in warehouses shifting various materials from place to place 1 cu yd at a time.

Development History

Prior to the development of a special-purpose machine known as the tractor shovel, a few firms built loading attachments for use on farm tractors. These scoops had limited mobility but were suitable for light material-handling tasks. One of the firms that manufactured such attachments was the W. M. Blair Manufacturing Company. In 1933, W. M. Blair was bought by an entrepreneur named Frank G. Hough. In 1939, for reasons not clear to those working with him, Hough decided to develop a specialized machine that could *only* be used for loading tasks but would be highly efficient at that function. There is no indication that Hough was spurred to undertake this task either by user requests for such a machine or user development activity.

Lowell Conrad, chief engineer at the Hough Company at that time, recalls that the development of the first tractor shovel was carried out by all the engineers then working at the company (only two or three!). The task took about six months and cost approximately $13,000 in parts and labor (engineers' wages were $200–$300 per month). Costs were kept down in part by obtaining heavy-duty tractor parts from International Harvester on consignment. Other parts were made by Lowell Conrad at night in a neighboring machine shop.

Commercialization

The tractor shovel that Hough and his engineers developed was called the Model HS and was first commercialized in 1939. Approximately 100 of the Model HS tractor shovels were sold from first commercialization until 1941 when production was stopped by wartime requirements.

In 1944 the Hough Company again began to produce tractor shovels, this time offering a small model (HA) suitable for unloading box cars. In 1946 the first competition appeared in the form of a three-wheeled tractor shovel produced by the Scoopmobile Company located in Oregon. In 1950 the Tractomotive Company of Ohio (later of Deerfield, Illinois) also produced a tractor shovel.

Side Lift Arm Linkage

The scoop of the original tractor shovel was attached to the tractor by a guide-frame linkage—the same sort of linkage used on a modern forklift truck. The linkage consists basically of two vertical rails attached to the tractor. The scoop is moved vertically up and down these rails. The side lift arm linkage that replaced it looks more like the linkage that attaches the blade to a modern bulldozer. In this linkage the scoop is attached to the tractor by two large horizontal beams that extend from the sides of the scoop to the sides of the tractor. The scoop is lifted by pivoting these beams around their point of attachment to the tractor.

The side lift arm linkage offers two great advantages to tractor shovel users. First, the pivoting motion that lifts the scoop also has a horizontal component that pushes the scoop slightly forward at the beginning of the lifting movement. This forward component of motion, called crowding, is very helpful when the operator is trying to scoop up hard-packed material. Second, the side lift arm linkage greatly reduces the overall height of the tractor shovel. In the older guide rail linkage, the guide rails were vertically fixed permanently, with their upper ends always at the height of the maximum vertical extension of the tractor shovel scoop.

Development History
Lowell Conrad, chief engineer of the Hough Company, conceived of, and developed, the side lift arm linkage innovation. Conrad recalled deciding to develop such a linkage as a result of a trip to a tractor dealer in Kansas City. The dealer told Conrad of the many underground mining operations in lead and limestone then existing around Kansas City. Existing tractor shovels could not be used in these mines because of the high clearance required by the guide frame linkage. The dealer emphatically told Conrad that he could sell many tractor shovels to these customers if the clearance problem could be eliminated.

Commercialization
In 1945 Hough was first to introduce side lift arm linkages on tractor shovels. Lowell Conrad estimates the total direct cost of the development work—given the staffing, salaries, and material costs of the day—at about $5000.

Side lift arm linkages were next commercialized by Scoopmobile on their Model C in 1948. Transco followed in 1950 as did Tractomotive.

Power Steering

The power steering systems used on tractor shovels are much like those used on cars and trucks. That is, they involve a hydraulic system that assists the steering efforts of the operator. Since the tractor shovel carries heavy loads on its front (steered) wheels, steering effort is high and the introduction of power steering reportedly increased tractor shovel productivity significantly.

Development History
In 1948 power steering was standard equipment on many types of heavy-duty construction equipment. The essential components of these systems were manufactured by firms such as Vickers and Garrison, specialists in hydraulic equipment. In 1947 Garrison came to the Hough Company and suggested they install power steering on their larger tractor shovels. They offered technical help and what was in essence a kit of components that could conveniently be adapted by Hough engineers to their equipment.

Commercialization
In 1947 Hough offered a kit based on Garrison components that users could install on their machines in the field. In 1948 Hough offered a factory-installed version of the Garrison system as an option on their four-wheeled drive (Model HF tractor shovel).

In 1948 the Dempster Company offered a Vickers hydraulic system on their Digster model; Scoopmobile offered power steering on its Model C.

Hydraulic Bucket Control

The scoop of a tractor shovel is emptied by rotating it until the open side of the scoop faces downward and the material carried simply falls out. Prior to hydraulic bucket controls, the bucket was attached to a pivot point below the bucket's center of gravity. On the release of a mechanical latch, the bucket simply flipped over in an uncontrolled manner and released the total contents. The hydraulic bucket control substituted a hydraulic piston and valve so that the operator could regulate the rotation of the bucket and thus dump a load partially or fully—a very useful capability. Hydraulic rotation of the bucket was also valuable in the digging of hard-packed material. An operator could wedge a lip of the scoop into the material and then rotate the bucket to break it loose.

Development History

The physical embodiment of the hydraulic bucket control was a hydraulic piston mounted between the bucket and a yoke fitted appropriately to the scoop lift linkage. It was developed by Lowell Conrad's engineering group at Hough in 1947. Since a source of hydraulic power was already on tractor shovels, development effort required to design and install the hydraulic bucket control—once the insight was available—was minimal in Lowell Conrad's recollection. Conrad estimates that it cost approximately $5000 in labor and materials to design and prototype the first hydraulic bucket control.

Commercialization

The hydraulic bucket control was first commercialized in 1947 on the Hough HLD Model. After Hough commercialized this innovation, other manufacturers of tractor shovels quickly imitated and offered functionally equivalent systems.

Fluid Transmission Coupling

As owners of manual transmission automobiles are aware, when a clutch slips, it rapidly wears out. Operators of tractor shovels found themselves often having to slip the clutch when they were trying to move heavy loads. As a result clutches on these machines often needed replacement after only a few weeks' use. The fluid transmission coupling is an innovation designed to solve this problem. In essence, it is a hydraulic component of the drive train that can absorb slip nondestructively. Tractor shovels equipped with it allow operators the same mechanical torque advantages formerly achieved by slipping the clutch, but without the associated clutch wear.

Development History

According to interviewees, frequent customer complaints and requests for replacement parts made all tractor shovel manufacturers very aware of the clutch wear-out problem. Since Scoopmobile engineers knew that fluid couplings had been successfully used to solve clutch-slippage problems on other types of construction equipment, it was natural for them to adopt the same solution.

Appropriate fluid couplings were available from suppliers as a standard component. Since Scoopmobile used Chrysler industrial engines in its tractor shovels, it was able to install Chrysler fluid couplings with a minimal development cost. (At this time, Scoopmobile's entire engineering department consisted of three people, at an average salary of $100 a week.)

Commercialization

In late 1948 Scoopmobile offered a fluid coupling on their Model C tractor shovel.

Planetary Final Drive

In the late 1940s tractor shovel manufacturers purchased heavy-duty truck axles from major suppliers and adapted these to their machines. These axles were not strong enough for some of the rough conditions encountered by tractor shovel users, and they often broke. The planetary final drive innovation was an ingenious idea that allowed manufacturers of tractor shovels to continue to use truck axles but to reduce the torque experienced by these and, thus, eliminate the breakage problem. This effect was achieved by installing a final gear reduction in the form of a planetary gear system in the hub of each driven wheel of the tractor shovel.

Development History

The principles of planetary gear systems have been well-known to engineers for many years. Harry Fielding, chief engineer of Scoopmobile, recalls that he developed the application of this principle to the problem of torque reduction in tractor shovel drive axles as a result of seeing an old steering wheel from an early Ford car in a junkyard. This steering wheel contained planetary drive-gear reduction in its hub. Fielding quickly developed planetary final drive for Scoopmobile tractor shovels and recalls that the direct cost of the development work as only several hundred dollars.

Commercialization

In late 1948 or early 1949 Scoopmobile was the first to commercialize planetary final drive on their Model CF tractor shovel. In 1953 Clark Equipment Company offered planetary final drive on their new line of tractor shovels (they made their entry into the business at this time). In 1953 Hough also offered planetary final drive.

Double-Acting Hydraulic Cylinders

The pistons of double-acting hydraulic cylinders can be moved both in and out by the application of hydraulic force. In contrast, the pistons of single-acting hydraulic cylinders can only be moved in one direction by hydraulic force. Until the innovation of double-acting cylinders, hydraulic pistons installed on tractor shovels were all of the single-acting variety. This meant, for example, that tractor shovel operators could lift the scoop of their machine with hydraulic force, but they were forced to rely on the weight of the scoop itself for any motion in a downward direction.

The introduction of double-acting cylinders gave operators increased control over the scoop of the tractor shovel and allowed them to apply increased downward force. This was very useful in many tasks—as in, for example, scraping ice from streets in winter.

Development History

Hough was the first company to introduce double-acting cylinders to tractor shovels. According to interviewees at that company, they were motivated by a particular sensitivity to the municipal market for street clearing. For this application it was quite clear that down pressure provided by the weight of the bucket only was not sufficient. It was also clear that increased down pressure would be useful in such tasks as breaking up the asphalt surface of roads. Cost of implementing the innovation was quite low because double-acting cylinders and the controls needed to operate them were standard products of the hydraulic supply industry. All that was needed was to install double-acting cylinders in the locations occupied by single-acting cylinders and to add appropriate reinforcement to affected linkage points.

Commercialization

Double-acting cylinders were first commercialized by Hough in 1948 on their Model

HM. The rest of the industry quickly followed and also commercialized the improvement.

Four-Wheel Drive

Four-wheel drive offers much the same advantages to users of tractor shovels as it does to the users of four-wheel drive trucks and similar equipment. In essence, it offers better traction, especially on rough terrain, and rough terrain was often encountered by tractor shovels being used on construction sites.

Development History

The development of a four-wheel-drive tractor shovel was technically straightforward—but quite costly. Although the engineering required was well understood and although the axles needed for four-wheel drive were available as standard components from manufacturers, much of the tractor shovel had to be redesigned in order to incorporate the four-wheel-drive feature. For example, many existing components had to be shifted to make room for the new drive train needed. Also, provision had to be made to allow the rear axle to oscillate so that it might keep better contact with the ground on rough terrain. Total direct cost to develop the four-wheel-drive tractor shovel is estimated by Lowell Conrad at $70,000.

Commercialization

In 1948 four-wheel drive was first offered commercially by Hough on the Model HF tractor shovel. In 1953 Scoopmobile followed. Today, four-wheel drive is a common feature of tractor shovels.

Torque Converter

A torque converter is a hydraulic mechanism used in transmissions that provides an effect equivalent to infinitely variable gearing within a narrow range. When used in the transmission of a tractor shovel, it enables operators to adjust the torque they apply to a task more precisely than can be done with manual transmission alone. Tractor shovels equipped with a torque converter need not also have a fluid coupling because the function of this device is also inherent in the torque converter. (See the earlier description of fluid coupling.)

Development History

Torque converters developed by Allis Chalmers had been used in construction machinery as early as 1947. Tractomotive, a manufacturer of tractor shovels, had a history of close association with Allis Chalmers, indeed, it was eventually acquired by that firm. If Tractomotive were already using Allis Chalmers' transmissions and engines (which seems likely, although we have no direct evidence), then adoption of the Allis Chalmers torque converter would have involved almost no development work on the part of Tractomotive. Design work necessary to integrate the Allis Chalmers torque converter with the Allis Chalmers engine and transmission would already have been done by that firm.

Commercialization

The torque converter was first offered as a commercial feature of tractor shovels by Tractomotive in 1951. In 1953 the Hough Company offered it on their Model HMC.

Articulation

Articulation like four-wheel drive involves major changes in the configuration of the tractor shovel. In effect, the entire machine is split in two at a point between the front and rear wheels and then reattached by means of a hinge. This hinge is controlled and flexed by means of hydraulic cylinders mounted on either side of it. Articulation eliminates the traditional steering system: The driver turns by bending the machine in the middle at the hinge point.

Articulation offers several advantages to the user. First, it provides a shorter turning radius. Second, it reduces maintenance costs and downtime related to, and caused by, steering mechanisms and their failure. Third, with articulation the rear wheels follow the same path as the front wheels, and both sets follow directly in the path of the front scoop, which can thus be used to clear the way.

Development History

Articulation was developed at the Scoopmobile Company. The direct motivation to the development work (according to Harry Fielding, chief engineer of that company) was inventory reduction rather than performance. Front and rear axles are the most expensive components of tractor shovels, and an articulated machine uses the same type of axle both front and rear. Articulation had been developed and patented prior to 1953 by a British manufacturer of construction equipment. Fielding claims Scoopmobile was unaware of the British patent when it did its own development work and only found out about it when conducting a patent search.

Commercialization

Scoopmobile was first to commercialize articulation on tractor shovels, introducing the innovation with their LD 10 model in 1953. Scoopmobile salesmen quickly became expert at graphically illustrating the advantages of articulation. Competitive machines would often embarrassingly fail to perform or mire down on customers' sites under conditions the Scoopmobile LD 10 proved able to handle. Despite this demonstrable superiority, competitors were relatively slow to adopt articulation. In part this was probably because articulated machines looked decidedly odd to the traditional eye; in part it was probably because Scoopmobile was a small regional company most of whose sales were restricted to the Pacific Northwest. Therefore, the advantages of articulation did not really offer a significant competitive threat in the major markets of other producers.

In 1962 the Euclid Company developed an articulated machine that used the basic method spelled out in the British patent. In 1964 Caterpillar, which did not begin to produce tractor shovels until 1959, introduced articulated machines, as did the Hough Company with their Model H120C. Today, all large tractor shovels use this innovation.

Power Shift Transmission

The power shift transmission is a form specially suited for the requirements of tractor shovel operation. All available gears are permanently in mesh, and each is mechanically linked to a separate clutch. Thus, if the transmission has four gears, it has four separate clutches. Selection of a gear in a power shift transmission simply involves engaging the proper clutch by the simple motion of a hand lever connected to a hydraulic actuator. All foot action on the part of the operator in the shifting process is eliminated.

On many tasks that tractor shovels perform, operators have to shift very frequently—as often as 10 times a minute. In tasks requiring frequent shifting, operator fatigue from that task significantly constrained tractor shovel productivity. This was graphically illustrated by the increased productivity shown by tractor shovels equipped with power shift transmissions.[1]

Development History

The first power shift transmission was developed by Clark Equipment Company in 1952 and 1953. Clark was a new entrant into the tractor shovel business, and they wanted the first models they offered to have a power shift transmission.

Lowell Conrad left Hough to join Clark and develop the power shift transmission; he estimates its development cost at $250,000.

Commercialization

Clark introduced the power shift on their new line of Michigan brand tractor shovels in 1954. This line of shovels was commercially very successful. Indeed, within two years Clark was selling as many tractor shovels as Hough—the firm that had traditionally dominated the industry. In 1955 Hough introduced a form of power shift transmission based on the GM Truckmatic drive in their HO model.

Note

1. "Construction Equipment: Ten Years of Change," *Engineering News Record* 170 (21 February 1963): 45.

DATA SET FOR ENGINEERING PLASTICS

The data set for engineering plastics examines innovations in engineering thermoplastic materials. Engineering plastic is the common term for materials that can compete with traditional engineering materials such as metals, wood, and glass from the standpoint of such properties as strength, temperature resistance, and ease of fabrication. Engineering thermoplastics are thus suitable for such applications as gears and motor housings—applications for which metals were traditionally used. Sample selection criteria for this data set will be found in chapter 3.

Polycarbonate Resins (Lexan)

Polycarbonate resins (better known in the United States under the General Electric trade name of Lexan) have several important properties that make them excellent engineering materials. They have high-impact strength, superior dimensional stability, transparency, and good electrical insulation qualities. They are also self-extinguishing—an important property where fire safety is at issue. These properties have made polycarbonate of great value in applications ranging from precision camera components (an application where dimensional stability is important) to football helmets (an application where toughness is critical).

Development History

Polycarbonate engineering plastics were developed independently by two firms: GE in the United States and Farben Fabriken Bayer AG of West Germany. Research work in both firms began about 1953. Commercialization (i.e., the sale of developmental quantities of the material manufactured in a pilot plant), however, was first achieved by GE; we, therefore, regard GE as the innovating firm.

Lexan polycarbonate resins were discovered by Dr. Dan Fox, a researcher at GE, through serendipity. In 1953 Fox joined a team of GE researchers working to develop an improved electrical insulation for magnet wire. Insulated magnet wire is a critical part of electrical motors and other apparatus manufactured by GE. Development of a thinner insulating material capable of withstanding high temperatures would allow the construction of better performing and smaller electrical motors and generators.

When Fox joined the research team, a family of polymeric materials had been discovered that had the proper flexibility, toughness, and resistance to high temperature, but all were somewhat degraded by water. Fox recalled some postdoctoral research he had conducted at the University of Oklahoma in which he had used guaiacol carbonate and found it extremely hydrolytically stable. He, therefore, decided to try to make a polymer based on this material. He experimented with bisphenol-A and started making a polymer by ester exchange with diphenyl carbonate. The resulting polymer was Lexan.

The polymer discovered by Dr. Fox was not considered the most promising route to a wire insulation of the desired properties and was put on the shelf until the completion of the wire insulation project in 1954. In 1954 the chemical development department of GE decided to make a polycarbonate resin to be used as a molding material—the first such material of its own GE would attempt to commercialize. Development of Lexan to the test-market stage was completed by 1958.[1]

Commercialization

General Electric began to test market polycarbonate resin in the United States in 1958; Farben Fabriken Bayer AG began to introduce the product in Europe in 1959. Given the independent research work of both firms in the polycarbonate field and given that both had independent patent positions, these two firms entered into an agreement not to contest each other's patents. Development cost for polycarbonate was approximately $10 million.[2] Sales of Lexan grew quickly: In 1960 approximately 500,000 lb were sold at a price of $1.50/lb; in 1969 approximately 31 million lb were sold at a price of $0.80/lb.[3]

Notes

1. National Academy of Sciences, *Applied Science and Technological Progress,* A Report to the Committee on Science and Astronautics, U.S. House of Representatives, GP-67-0399 (Washington, D.C.: U.S. Government Printing Office, June 1967), 35–37.

2. Author's estimate based on information supplied by Dr. Dan Fox, General Electric, Schenectady, N.Y.

3. Dirk Oosterhof, *Chemical Economics Handbook: Plastics and Resins,* 580.1120A. (Menlo Park, Calif.: Stanford Research Institute, March 1970), sections F and I.

Acetal Homopolymer Resins (Delrin)

In the United States, acetal homopolymer is probably best known by the Du Pont brand name, Delrin. Delrin is a nontransparent engineering plastic. The plastic has a regular structure and high crystallinity that gives parts made from Delrin high strength and rigidity, excellent dimensional stability, and resilience over a wide range of service temperatures and humidities and a wide range of solvent exposures. In addition, the plastic has excellent frictional properties, which allows the plastic to serve as a good mechanical bearing.

Developmental History

Delrin is a polymer of formaldehyde. The existence of formaldehyde polymers had been known since before the 1920s. Interest in them was limited, however, because they were felt to be inherently unstable. In 1947 A. Barkdoll, Jr., of Du Pont's chemical department, began to study formaldehyde monomer and discovered that the pure monomer sometimes spontaneously polymerized into a polymer with attractive properties. In the 1949–50 period a project under R. MacDonald was begun that was explicitly devoted to the development of formaldehyde polymers. In late 1952 the research looked so promising that the polychemicals department (one of Du Pont's major industrial subdivisions) put 60 men to work on the project. Up to 1960, Delrin R & D expenses were approximately $27 million.[1]

Commercialization

Du Pont invested $15 million in a plant at Parkersburg, West Virginia, capable of producing about 15 million lb of Delrin annually, and introduced it commercially in 1960. Du Pont is the sole producer of acetal homopolymer. Its high expectations for the commercial success of the plastic were not fully met, however, owing to the later commercialization of an acetal copolymer called Celcon by Celanese in 1963. (See discussion of acetal copolymer below for details.) In 1974 approximately 70 million lb of acetal resins were produced with one quarter of this being acetal homopolymer resin (Delrin). The sales price of Delrin was $0.95/lb in 1960 when the product was introduced and was $0.80/lb in 1974.[2]

Notes

1. Herbert Solow, "Delrin: du Pont's Challenge to Metals," *Fortune* 60, no.2 (August 1959): 116–19.
2. *Chemical Economics Handbook: Plastics and Resins,* 580.0121 (Menlo Park, Calif.: Stanford Research Institute, October 1975), sections B and E. The percentage of acetal resin production consisting of acetal homopolymers was obtained from *The Kline Guide to the Plastics Industry,* ed. James A. Rauch (Fairfield, N.J.: Charles H. Kline & Co., 1978), 56.

Acetal Copolymer (Celcon)

Acetal copolymer is a polymer of formaldehyde. It has slightly better properties with respect to heat and solvent resistance than does the closely related acetal homopolymer (Delrin) (see preceding entry).

Development History

Celanese Corporation is a basic producer of formaldehyde—the feed stock for acetal homopolymers and copolymers. Du Pont's demonstration in 1960 that a formaldehyde polymer with commercially attractive properties could be produced induced Celanese to accelerate its research in the area. Intensive research and development at Celanese—with Drs. Brown, Barting, and Walling essential to the effort—allowed Celanese to produce pilot-plant quantities of Celcon in April 1961 in Clarkwood, Texas.[1] A long and difficult court battle then ensued between Du Pont and Celanese over patents—Celanese finally won the case.

Commercialization

Du Pont's experience with the introduction of Delrin allowed Celanese to proceed to full-scale production rapidly, confident of Celcon's commercial attractiveness. Celanese's first plant for the production of Celcon went on-stream at Bishop, Texas, in January 1962.[2] Approximately 125 million lb of acetal resins were produced in 1974 by Du Pont and Celanese.[3] Approximately 35 million lb of this was exported; approximately 90 million lb was consumed in the U.S. market. Approximately three quarters of the total volume was acetal copolymer.

Notes

1. "Celanese Celcon Plant On Stream," *Modern Plastics,* 39, no. 6 (February 1962): 45.
2. *Chemical Economics Handbook: Plastics and Resins,* 580.0121A (Menlo Park, Calif.: Stanford Research Institute, October 1975), 580.
3. James A. Rauch, ed., *The Kline Guide to the Plastics Industry* (Fairfield, N.J.: Charles H. Kline & Co., 1978), 56.

Polysulfone Resin

Polysulfone is composed of phenylene units linked by three different chemical groups—isopropylidene, ether, and sulfone—each contributing properties to the polymer that are conventionally obtained through stabilizers or other additives. Polysulfone is a strong, rigid thermoplastic that can operate in high-temperature environments. The heat distortion temperature of polysulfone at 264 lb/sq in. is 345°F. Polysulfone also has a high rating for self-extinguishability. Its combination of characteristics has made it important in applications involving long-term service at high temperatures under load.

Development History

Polysulfone was developed by Dr. H. Farnum and Dr. Robert Johnson, researchers at Union Carbide Corporation. They were searching for a thermoplastic material that could withstand high temperatures—motivated by a finding of Union Carbide's marketing research that there was a market for an engineering plastic that was stable at up to 175°C. Farnum and Johnson had a good theoretical sense that phenylene units in a polymer chain would exhibit satisfactory high-temperature characteristics. They synthesized and analyzed 40 to 50 possible molecular structures before deciding on the presently commercialized polysulfone.

Commercialization

Polysulfone was commercialized by Union Carbide in 1965. Union Carbide remains the sole producer. Production began in a plant constructed in Marietta, Ohio, capable of producing 10 million lb/year. The initial price was $1.00/lb.[1] In 1976 production was approximately 12 million lb with a market price of approximately $22 million.[2]

Notes

1. "New Engineering Thermoplastic—Polysulfone," *Modern Plastics* 42, no. 9 (May 1965): 87–89, 196.

2. James A. Rauch, ed., *The Kline Guide to the Plastics Industry* (Fairfield, N.J.: Charles H. Kline & Co., 1978), 55.

Modified Polyphenylene Oxide (Noryl)

Noryl is a thermoplastic with outstanding dimensional stability at elevated temperatures. It is hydrolytically stable and has both excellent dielectric properties and chemical resistance. It is widely used in automotive and appliance applications.

Development History

Noryl was developed by General Electric researchers and is a commercially successful modification of an earlier GE polymer, polyphenylene oxide (PPO). Research that led to PPO was begun by A. S. Hay in the mid-1950s in the GE Research Laboratory in Schenectady, New York, where he was working on the oxidation of organic compounds, particularly phenols. According to Hay his research had no application in mind. In 1956 he discovered polymerization of phenols by oxidative coupling—a new chemical technique for synthesizing plastics.

Because of GE's then-ongoing work on Lexan, Hay's discovery was not developed further until 1960. In 1960 an effort was begun by J. R. Elliot of the chemical development operation to find an economical means of producing the product commercially.

In 1965 a pilot plant was ready to produce PPO and the product was commercialized. Various applications such as surgical instruments and appliance parts were made with PPO, but by 1957 it was clear that PPO was a commercial failure. It had proven hard for molders to process because it required specialized molding machinery and a drying step.[1] As the commercial failure of PPO became clear, work was begun to modify the material. Noryl was the result of this research: an alloyed PPO that is easy to process.

Commercialization

Noryl was introduced commercially in 1968. Its sole manufacturer is GE. In 1976 approximately 95 million lb were sold with a sales price of approximately $90 million.[2]

Notes

1. National Academy of Sciences, *Applied Science and Technological Process,* A Report to the Committee on Science and Astronautics, U.S. House of Representatives, GP-67-0399 (Washington, D.C.: U.S. Government Printing Office, June 1967), 37.

2. James A. Rauch, ed., *The Kline Guide to the Plastics Industry* (Fairfield, N.J.: Charles H. Kline & Co., 1978), 55.

DATA SET FOR PLASTICS ADDITIVES

The data set for plastics additives contains information on innovations for two types: plasticizers and ultraviolet stabilizers. Specific sample selection criteria will be found in chapter 3. Data on all innovations listed in that sample will be found here, with the exception of four coded NA in chapter 3, Table 3–8. In these instances we could not find needed data after diligent research.

A plasticizer is a material that is incorporated mechanically into a plastic to increase its flexibility and workability. Without plasticizers polymers such as polyvinyl chloride (PVC) are hard and brittle; with plasticizers they become softer, more flexible, and easier to process. A UV stabilizer is a compound that protects plastic from the degrading effects of UV light. In the absence of UV stabilizers, polymers exposed to UV show loss of physical properties and discoloration, often accompanied by surface crazing (the formation of many fine cracks on the surface of a plastic), embrittlement, and chalking.

In chemistry-related fields, *Chemical Abstracts* is a major resource for the type of study conducted here. As a cross-check on our findings regarding the first to develop innovations researched, we searched *Chemical Abstracts* for several years prior to the date of each innovation to check for any application earlier than that we had identified by interview. Market size and development cost estimates given by interviewees in some of the innovation histories that follow are approximate and informal. Development cost figures, when provided, refer to direct costs only.

PLASTICIZERS

Butyl benzyl phthalate

Butyl benzyl phthalate is a plasticizer used in polyvinyl chloride (PVC). Its main advantage over earlier plasticizers used with PVC is that it fuses with PVC resins at a much lower temperature. This meant that PVC could be processed on standard rubber-processing machines that operate at 110° to 130°C rather than on special PVC milling machines that operate at 150° to 160°C. Rubber processors were thus able to change over from rubber to PVC processing without purchasing new machinery.

Development History

In the 1930s Bayer took out a patent for butyl benzyl phthalate as a plasticizer for cellulose nitrate. A Monsanto research team under the leadership of Joe Darby (manager of the plastic applications laboratory) reasoned that butyl benzyl phthalate might also be a good plasticizer for PVC. When they discovered the attractively low temperature at which butyl benzyl phthalate could be blended with PVC, they took it to the marketing department to be considered for commercialization. Monsanto also took out process patents in 1946 to cover improved means of producing the plasticizer.

Commercialization

Monsanto introduced butyl benzyl phthalate in 1946 and marketed it to rubber and PVC products. In 1974 approximately 80 million lb of the plasticizer was sold at a market price of approximately $28 million. Monsanto interviewees estimate the cost of the development of the plasticizer at approximately $500,000 to $1 million.

2 ethyl hexyl di phenyl phosphate

2 ethyl hexyl di phenyl phosphate is a plasticizer that equaled the desirable cold flex and volatility characteristics of competitive plasticizers and also imparted flame retardancy to plasticized polyvinyl chloride (PVC). In addition, the plasticizer is nontoxic and has Food and Drug Administration (FDA) approval for use in food-packaging films, a major application.[1]

Development History

In 1945 Union Carbide introduced a PVC insulation material that was plasticized with di octyl phthalate. This product showed excellent cold flex characteristics and Monsanto was anxious to match it or exceed it. Monsanto, therefore, began looking for a low toxicity, flameproof plasticizer that would be equivalent to di octyl phthalate in cold flex and volatility. The product they developed succeeded in combining the good cold flex properties of the di octyl phthalate with the excellent flame-proofing and solvating properties of the aryl phosphates to give a product that generally has many of the best properties of both compounds.

Commercialization

2 ethyl hexyl di phenyl phosphate was commercialized in 1947 by Monsanto. There is no competitive producer of this product in the United States, but Monsanto itself has since developed a second product of this type—iso decyl di phenyl phosphate—that provides even better cold flex and volatility products. These two plasticizers have tended to replace tri alkyl phosphates and tri butyl phthalates in the marketplace.

Note

1. J. Kern Sears and Joseph R. Darby, *The Technology of Plasticizers* (New York: Wiley, 1982).

Citroflex Plasticizers

Citroflex-type plasticizers are used for vinyls and cellulosics. Examples of such plasticizers are: tri butyl citrate and acetyl di butyl citrate. The principal advantage these plasticizers offered over existing plasticizers in 1957, the year of their introduction, was FDA approval to use them to plasticize films used to package oily and greasy foods. Phthalate plasticizers, by way of contrast, only have FDA approval for use in films to package high-water–content foods.

Development History

Pfizer Corporation, a producer of citric acid, was anxious to develop uses for that acid. Also, it was anxious to move out of the phthalic ester plasticizer business because it was basic in neither the anhydride nor the alcohol components of this product. Both problems were addressed by a research project that attempted to develop the citroflex plasticizers. The work proceeded from 1953 to 1957 and resulted in the development of a range of citroflex esters. The costs of development per ester were estimated by a participant in the research project as approximately $125,000. Toxicological testing of each ester was estimated at $150,000 to $250,000.

Commercialization
Pfizer introduced the citroflex-type plasticizers in 1957. It is the only supplier of these esters to the present day. Dollar volume of sales in 1974 is held in confidence by Pfizer.

Note
1. Arnold L. Baseman, "A Compounder's Guide to . . . Plasticizers '65," *Plastics Technology* 11, no. 10 (October 1965): 37–44.

Di N undecyl phthalate Plasticizers

The plasticizer di N undecyl phthalate is one of two long alkyl chain phthalates that are widely used for plasticizing PVC cable wiring designed to operate at 75° C and above. These plasticizers are superior to the branched tri decyl phthalates formerly used for this purpose in three major ways: (1) they provide a superior cold flex at low temperatures; (2) they fuse with PVC at a lower temperature than the earlier plasticizers, which eases processing requirements; (3) they are more compatible with PVC than were earlier plasticizers, which tended to bleed at flexing points in the cable.

Development History
The need for a good, highly compatible plasticizer suitable for use with PVC subjected to high temperature was widely known within the plasticizer industry. From 1967 to 1970 the Monsanto plasticizer applications laboratories worked to develop such a product. The approximate cost of the research over the three-year development period was between $500,000 and $1 million.

Commercialization
The di N undecyl phthalate plasticizers were introduced commercially by Monsanto in the early 1970s. Monsanto has requested that its sales and market share be kept confidential.

Tri melitate Plasticizers

Tri melitate plasticizers are primarily used to plasticize polyvinyl chloride (PVC). The product had a lower volatility than previous plasticizers and thus reduced application problems such as the fogging of automobile windows by plasticizers volatilized from automobile parts. The new plasticizer also offered improved cold flex properties.

Development History
The precursor to tri melitate plasticizer was provided by the Amoco Company in 1961. In that year Amoco built a plant capable of producing 2 million lb of tri mellitic anhydride per year. As part of the effort to develop markets for its product, Amoco sent samples to at least three chemical companies suggesting that the product might be a component of a good plasticizer. Monsanto, Pfizer, and W. R. Grace all added alcohol to the sample to make tri mellitic esters and tested these as plasticizers.

Commercialization
On observing the good properties of tri mellitic esters, Monsanto, Pfizer, and W. R. Grace all commercialized the chemical as a plasticizer in 1961. In 1974 approximately 23 million lb of the plasticizer were sold by all three companies together.

Some \$100,000 to \$250,000 was spent on developing the new stabilizer. The market size for the innovation about 1970 was approximately \$300,000 per year.

ULTRAVIOLET STABILIZERS

2:4 dihydroxy benzophenone

Polymers such as polystyrene and polyvinyl chloride are particularly susceptible to UV light in the 290 to 400 nm region. This UV light causes breakdown of the chemical bonds in the plastic, which in turn causes the plastic to deteriorate. Unprotected plastics may, for example, discolor and become brittle in the sun.

Benzophenone compounds were the first effective UV stabilizers on the marketplace. Typically, a PVC stabilized with a benzophenone compound would have a useful life one hundred times greater than that of unstabilized polyvinyl chloride in outdoor applications.

Development History

In the late 1930s or early 1940s Dow Chemical Company researchers patented 2 hydroxy 5 chloro benzophenone as a UV stabilizer. This compound proved to be only a weak absorber of UV light and gave a strong yellow color when used. It was not successfully developed commercially by Dow as a UV stabilizer for plastics.

I. G. Farben, a German firm, manufactured plastics of all sorts as well as a wide range of additives. During World War II, assets of I.G. Farben in the United States were sequestered by the U.S. Government. Employees of General Analine and Film (GAF), operator of the sequestered assets, found a considerable amount of data in the former I.G. Farben laboratories on UV-active compounds. A team of GAF researchers, Dr. F. Newmann, Dr. McKay, and Dr. Albert Strobel, began a program to develop a range of stabilizers for cellulose acetate. The team investigated a whole range of compounds found to be UV-active and determined that 2:4 dihydroxy benzophenone was an excellent stabilizer. (According to Dr. Strobel, at the time the GAF team did not know of Dow Chemical Company's earlier work on halogenated 2:5 benzephenones.)

Commercialization

General Analine and Film began commercial production of its first UV stabilizer sometime between 1948 and 1950. The total market volume of benzophenone-type stabilizers was perhaps \$10 to \$15 million per year in 1973. Dr. Strobel estimates that the cost of development, including monies spent on developing a production process, at about \$1 million. The second producer of benzophenone UV stabilizers was American Cynamid.

Ethyl-2-cyano 3:3 diphenylacrylate

Acrylate UV light stabilizers are used in polar polymeric systems. Acrylates have replaced benzophenones in applications where colorlessness is critical. Unlike benzophenones, they do not produce a yellowish color over time.

Development History
The acrylate group of ultraviolet stabilizers was developed by GAF as part of a program to produce new UV stabilizer systems for the plastics business. The research began in 1947 and continued until 1952.

The idea for the acrylate group of stabilizers was developed by Dr. Albert Strobel of GAF through his knowledge of dye chemistry. As a result of experiments on dyes, he knew that hydroxy benzaldehyde and cyanoethyl acetate reacted to give a dye that was extremely light-stable and a powerful absorber of UV light. Dr. Strobel realized that if he could alter the molecule to remove the dye properties while retaining the light stability and the UV-absorbing properties, he would have a good UV stabilizer. Through extensive research he was able to achieve this result.

Commercialization
Acrylate UV stabilizers were commercialized in 1952 by GAF. On the whole they have been very successful commercially, replacing benzophenones in applications where colorlessness is critical in polar polymers. In general-purpose applications, however, they have not replaced benzophenones because they are more expensive on a cost-performance basis. In 1973 the sales volume of acrylate stabilizers—approximately 1/3 of 1 million lb/year—made them probably the third largest UV stabilizer group in volume terms. (In that year benzophenones sold approximately 2 million lb and benzotriazoles sold approximately 500,000 to 1 million lb. The market value of 1/3 of 1 million lb of acrylate stabilizer is approximately $2 to $3 million.)

2 hydroxy 4 dodecyloxy benzophenone

Polyolefins are a commercially important plastic that absorbs ultraviolet light strongly in the region of 300 Å to 310 Å. The absorption of this energy causes the polymer chains to break; this in turn causes surface crazing, embrittlement, chalking, discoloration, and loss of physical properties such as impact and tensile strength. 2H 4D benzophenone was the first colorless stabilizer that was compatible with polyolefins.[1] Prior to the development of this stabilizer only filled pigmented polyolefins could be used out-of-doors and even these were not as stable as was desired. Currently about 70% of all UV plastic stabilizers are used in polyolefins.

Major uses of UV-stabilized polypropylenes are in fibers for carpeting and upholstery and the like. Typically, the addition of a stabilizer extends the life of polyolefin 10 to 20 times in outside duty.

Development History
Most experts in the field of UV stabilizers were aware that a 4 long-chain alkoxy-substituted hydroxy benzophenone would be a good stabilizer for polyolefins. The major problem in producing such a compound was to economically alkylate the 4-position OH group while leaving the 2 OH group unreacted. The problem was solved by researchers at Eastman Kodak, who in 1958 took out a patent for 2 hydroxy 4 dodecyloxy benzophenone.[2] Dr. G. Arick of Eastman states that this product was developed for internal use and that Eastman Kodak is a major user of its own product.

Commercialization
Eastman Kodak first introduced the innovation commercially in 1960. Although they do sell the product, they also consume a substantial amount in-house. Eastman

Kodak's first competitor was American Cyanamid, who in 1961 introduced 4 octoxy 2 hydroxy benzophenone.

Notes

1. Note that polar benzophenones had been produced by GAF in the early 1950s, including 2 hydroxy 4 methoxy benzophenone. This was an excellent stabilizer, but it was incompatible with polyolefins.

2. U.S. Patent No. 2,861,053.

Nickel Complexes

Nickel complexes are used primarily to stabilize polyolefin polymers with respect to ultraviolet light. They have a significant advantage over previously available stabilizers for this purpose in that they simultaneously provide a site to which dye can adhere. Use of nickel complexes thus made it possible for the first time to dye polyolefins in strong colors. (Untreated polyolefins are not sufficiently polar to accept conventional dispersion dyes.)

Nickel complexes are synergistic in their effect with benzophenone UV stabilizers. When used together, these two compounds give 4 to 10 times the protection provided by either stabilizer used alone.[1] Although benzophenones are strong absorbers of UV light and thus protect the polymer, the nickel complexes act by a different mechanism. They act to quench groups on the polymer chain that are excited by UV light, thus preventing breakage of the molecule.

Development History

The nickel complex UV stabilizers were developed by Ferro Corporation. Ferro is a company that specializes in producing additives for plastics and paints. They do not produce the actual polymers themselves.

In the late 1950s Ferro engaged in a research effort to find additives in the polypropylene field in order to increase their product range and break into a new market. A. M. Nicholson of Ferro's research department headed a team that investigated 1400 different compounds for their properties with respect to polypropylene. The discovery that certain nickel compounds acted as dye sites was serendipitous. Ferro Corporation patented the most promising nickel compounds in 1960.[2]

Commercialization

Ferro Corporation commercialized the nickel complexes for use as stabilizers and dye sites in polyolefins in 1962. Ferro's first competitor was American Cyanamid.

Notes

1. Arnold L. Baseman, "UV Stabilizers for Plastics," *Plastics Technology* 10, no. 4 (April 1964): 30–35.

2. U.S. Patent No. 2,971,940 and No. 2,971,941. These patents, granted to A. M. Nicholson, C. H. Fucksman, and S. B. Elliot, were assigned to Ferro Corporation.

P methoxy benzylidene malonic acid dimethyl esters

P methoxy benzylidene malonic acid dimethyl ester ultraviolet light stabilizers were designed to compete with, and to be superior to, the acrylate stabilizers. They absorb UV in the shortwave-length region of the spectrum and thus give a high degree of protection to polar plastics such as polystyrene and polyurethane. The intrinsic light stability of the compounds and their absorptions on the low end of the spectrum give a colorless product that shows little yellowing owing to side reactions such as can occur with many of the benzophenone compounds.

Development History

P methoxy benzylidene malonic acid dimethyl ester UV light stabilizers were developed at American Cyanamid. This firm has an ongoing program to develop new and improved stabilizers for polypropylene and other polymers. A major part of this work consists of screening UV light-active compounds in order to determine their efficiency as stabilizers and their suitability in plastics systems.

Dr. Suret Susi, and American Cyanamid researcher, became aware through a survey of patent literature that the benzylidene types of materials might possibly be suitable UV stabilizers. He then prepared the p methoxy benzylidene malonic acid dimethyl ester compound and evaluated its properties as a stabilizer. A patent was applied for in April 1964.

Commercialization

Commercial production of p methoxy benzylidene malonic acid dimethyl esters began in 1964. Dr. Susi estimates that approximately $100,000 to $250,000 was spent on developing the new stabilizers.

2:4 di t butyl phenyl 3:5 di t butyl phenyl 4 hydroxy benzoates

2:4 di t butyl phenyl 3:5 di t butyl phenyl 4 hydroxy benzoates are UV stabilizers for polypropylene. They protect the polymer by absorbing ultraviolet light as does benzophenone, but they also react to stabilize polypropylene by a method that is not entirely understood. As a result this compound is claimed both to give enhanced life to polypropylene when judged relative to competitive UV stabilizers and to act as an antioxidant as well.

Development History

The UV stabilizer 2:4 di t butyl phenyl 3:5 di t butyl phenyl 4 hydroxy benzoate was developed by Dr. E. J. Smutney of Shell Corporation. Shell was a major manufacturer of polypropylene and wished to develop a good UV stabilizer in order to extend its market. In 1963 Dr. Smutney took out a patent on the innovative UV stabilizer.[1]

Commercialization

Shell Corporation did not commercialize the stabilizer. Although it knew it was a good product, it apparently could not produce the substituted acid precursor cheaply enough to make it a commercial success. Shell, therefore, approached several firms, including Ferro Corporation, with the product.

Ferro, a manufacturer of plastic additives, took a license from Shell to produce the innovation and developed an economical way of preparing the precursor acid for the stabilizer. This was patented[2] and allowed Ferro to produce the product at a commercially acceptable cost. Bill Meek of Ferro's research department states that he found a brief reference in a British journal to a carboxylic acid preparation that worked marginally well. He then improved the process by use of superior solvent chemistry. Ferro commercially introduced the product in December 1973. At present there is no competing commercial product. The cost to develop the product was approximately $25,000 to

$100,000. The cost to develop the process was approximately $100,000. The innovation has had some success in replacing benzophenones and benzotriazoles.[3]

Notes

1. U.S. Patent No. 3,112,338.
2. U.S. Patent No. 3,825,593.
3. Stephen C. Stinson, "Chemicals and Additives Today: The Pace of Development Quickens," *Plastics Technology* 18, no. 7 (July 1972): 35–49.

Zinc Oxide and Zinc diethyl dithio Carbamates

Zinc oxide and zinc diethyl dithio carbamate stabilizers are used in heavily pigmented plastic products. These products benefit only slightly from organic ultraviolet stabilizers since degradation occurs only at the surface of the product where there is only a small concentration of the stabilizer. The diethyl dithio carbamate acts synergistically with the zinc oxide giving a performance that is equal to, or better than, the performance provided by organic UV stabilizers at a lesser cost. Manufacturers claim that the life of pigmented plastics can be increased by 2 to 10 times when compared to plastics using titanium dioxide pigment and organic stabilizers.[1]

Development History

The zinc oxide and zinc diethyl dithio carbamate stabilizers were developed by two researchers at the firm of Debell and Richardson under the sponsorship of the International Lead and Zinc Research Organization. Stan Margosiak and Barry Baum of Debell and Richardson knew that zinc oxide had been used as a stabilizer for paints and that zinc diethyl dithio carbamate had been known to have antioxidant properties in rubber. This lead them to experiment with the effects of these two compounds in plastics. Debell and Richardson had been working on zinc stabilizer systems since 1967. A total cost of this particular project was about $100,000 to $200,000. The major costs involved testing the stabilized plastic systems.

Commercialization

The International Lead and Zinc Research Organization made the research results on the UV stabilizing properties of zinc compounds available to all its members. The compounds required were already produced by some members: New Jersey Zinc Company in Bethlehem, Pennsylvania, produces zinc oxide; zinc diethyl dithio carbamate stabilizers are produced by R. T. Vandervuilt Company in Norwalk, Connecticut. Thus, no commercialization expense had to be incurred by these firms. The value of the compounds in the UV stabilization of plastics was simply publicized to the plastics industry.[2]

Notes

1. Robert E. Hunt, "Chemicals and Additives '70," *Plastics Technology* 16, no. 7 (July 1970): 39–45.
2. D. S. Carr et al., "Zinc Oxide Stabilization of PP Against Weathering," *Modern Plastics* 47, no. 5 (May 1970): 114–18; D. S. Carr et al., "UV Stabilization of Zinc Oxide with Thermoplastics," *Modern Plastics* 48, no. 10 (October 1971): 160–61.

INDEX

N.B.: Page numbers in italics refer to tables and figures.